Literatures, Cultures, and the Environment

Series Editor
Ursula K. Heise
Department of English
University of California
Los Angeles, CA, USA

Literatures, Cultures, and the Environment focuses on new research in the Environmental Humanities, particularly work with a rhetorical or literary dimension. Books in this series explore how ideas of nature and environmental concerns are expressed in different cultural contexts and at different historical moments. They investigate how cultural assumptions and practices, as well as social structures and institutions, shape conceptions of nature, the natural, species boundaries, uses of plants, animals and natural resources, the human body in its environmental dimensions, environmental health and illness, and relations between nature and technology. In turn, the series makes visible how concepts of nature and forms of environmentalist thought and representation arise from the confluence of a community's ecological and social conditions with its cultural assumptions, perceptions, and institutions.

More information about this series at
http://www.palgrave.com/gp/series/14818

Begoña Simal-González

Ecocriticism and Asian American Literature

Gold Mountains, Weedflowers and Murky Globes

Begoña Simal-González
Universidade de Coruña
A Coruña, Galicia, Spain

Literatures, Cultures, and the Environment
ISBN 978-3-030-35617-0   ISBN 978-3-030-35618-7   (eBook)
https://doi.org/10.1007/978-3-030-35618-7

This Palgrave Macmillan imprint is published by the registered company Springer Nature Switzerland AG.
The registered company address is: Gewerbestrasse 11, 6330 Cham, Switzerland

To my father

ACKNOWLEDGMENTS

A book like this, which has taken ten years to grow from the apparently insignificant seed of an idea to the many-branched tree that we now see, is necessarily the product of stamina and love. If the fragile plant had not been watered by the generosity of (local and global) colleagues, it would have withered. I want to thank my ecocritical sisters and brothers, who, knowingly or unknowingly, inspired me to write this book (Margarita Carretero, Carmen Flys, Ursula Heise, Scott Slovic, and Molly Wallace). My thanks also go to those scholars who first encouraged me to carry out research on Asian American literature in the 1990s (King-kok Cheung, Rocio Davis, Sämi Ludwig, and Sau-ling Wong) and to the writers whose personal support made my research life so much easier (Maxine Hong Kingston, Shawn Wong, and Karen Tei Yamashita); to my colleagues at the American Studies Research Group (CLEU) and the Department of English, at the Universidade da Coruña; to my old friends from the Spanish Association for American Studies and from various research projects, who kept me going (Francisco Collado, Constante González Groba, Carmen Manuel, Isabel Durán, Mar Gallego, Cristina Alsina, Jesús Benito, Ana Manzanas, Antonieta Oliver-Rotger, José Liste, Martín Urdiales, Aitor Ibarrola, etc.). I am also grateful to Aoileann Lyons, who helped me with some stubborn English collocations, and to Emma Lezcano, who held my hand when it started to tremble.

For the groundwork and early stages of this book, I am indebted to the Regional Government of Galicia (Xunta de Galicia), for its generous funding of the research project *Human Ecology* (ref. PGIDIT07PXIB104255PR). The last leg of this publishing journey was made possible thanks to the funding received from the Spanish Research Agency (Agencia Estatal de Investigación) for the project "Literature and Globalization" (ref. FFI2015-66767-P, AEI/FEDER, UE). I also want to thank the anonymous readers of the early versions of the manuscript, whose insights and recommendations proved invaluable.

Water, in the form of academic help and institutional funding, was not the only reason why this tree of a book continued to grow. If it had not been for the warmth of family and friends, it would have remained stunted. I want to thank my godchildren (Irene and Carlos), who made me feel like a child again, and my closest friends (Natalia, Jaime, José, and Raquel), who kept me going when life seemed harder than usual. I will never forget the love of my mother, my sister, and especially "o meu home," Emilio, in these last months; thank you for being there during "la noche oscura del alma." Last but not least, I want to express my eternal gratitude to my father, who first taught me to discover the beauty and mystery of nature. Deep in my heart, I know he has found his particular heaven sitting by the Mocha de Polledo, the ancient tree at the center of the universe: the tiny village of Gramedo.

A Coruña
December 2018

UNIÓN EUROPEA
FONDO EUROPEO DE
DESARROLLO REGIONAL

AGENCIA
ESTATAL DE
INVESTIGACIÓN

XUNTA DE GALICIA
CONSELLERÍA DE INNOVACIÓN
E INDUSTRIA
Dirección Xeral de Investigación,
Desenvolvemento e Innovación

GOBIERNO
DE ESPAÑA

MINISTERIO
DE CIENCIA, INNOVACIÓN
Y UNIVERSIDADES

AGENCIA
ESTATAL DE
INVESTIGACIÓN

Praise for *Ecocriticism and Asian American Literature*

"Nothing could be further from the 'add-ecocriticism-and-stir' approach to literary studies than Begoña Simal-González's engaging, erudite, and utterly unpredictable book. Rereading key works of Chinese American and Japanese American literature through the twin lenses of critical race theory and ecocriticism, Simal-González excavates time- and place-specific botanical and zoological imaginaries deeply embedded in texts that are not explicitly about nature. Though I've been reading and teaching works by Sui Sin Far, Maxine Hong Kingston, and Karen Tei Yamashita for decades, Simal-González's theoretically informed close readings, which tease out the 'environmental unconscious,' made me see those works anew. The magisterial scope of this study, which reaches back to the 1880s and goes as far forward as the 2010s, allows Simal-González to show a wide range of literary deployments of nature, from Sui Sin Far's anti-racist plant and animal imagery to Chinese railroad and agricultural workers' imagined community with the land they transformed through their labor."

—Dominika Ferens, Associate Professor, *University of Wroclaw, Poland*, and author of *Edith and Winnifred Eaton* (2002)

"A landmark text, uniquely comprehensive in its scope, *Ecocriticism and Asian American Literature* engages a conversation long overdue—transforming both ecocriticism and Asian American literary studies in the process. Begoña Simal-González's thoughtful approach returns us to such issues as farming, internment, cultural nationalism, and transnationalism and diaspora, resituating these in a robustly historical environmental context. Essential reading for those interested in environmental justice in American literature."

—Molly Wallace, Associate Professor, *Queen's University, Canada*, and author of *Risk Criticism* (2016)

"Begoña Simal-González's book on ecocriticism is a most welcome addition to Asian American literary studies. All the writers included in the volume are not only unique voices that deserve a formal hearing but also pioneers in ecological interventions, as Professor Simal-González so persuasively demonstrated in her book."

—King-Kok Cheung, Professor of English and Asian American Studies, *University of California, Los Angeles, USA*, and author of *Chinese American Literature without Borders* (2016)

CONTENTS

LIST OF FIGURES

CHAPTER 1

Introduction

Let me tell you a story about black birds. In 2002, an oil tanker misleadingly called *Prestige* sank just a few miles off the Galician coast after a series of fatally wrongheaded technical and political decisions. The massive oil spill and environmental catastrophe that ensued mobilized activists and volunteers from all over Spain like never before.[1] The initial response of horror and stupefaction quickly turned to indignation and defiance. Cleanup teams worked for months to salvage and restore whatever they could from underneath the *marea negra* (black tide) smothering the coasts of Galicia and Asturias, while thousands of infuriated citizens joined environmental and social activists behind the rallying cry of *nunca máis* ("never again"). Never again would we be able to look in the other direction; the iconic image of seabirds coated in black, sticky tar had been seared into our minds forever. Ecological awareness was here to stay, not only on the streets but also in the academic circles in which I was just beginning to move.

Soon after the *Prestige* disaster, one of the pioneers of ecocriticism in Spain, Carmen Flys, approached me to ask if I knew of any Asian American texts that would be amenable to an ecocritical reading. At the time, only a few examples came to mind: the first novels of Karen Tei Yamashita and Ruth Ozeki, and a handful of Asian American poets. In fact, a quick perusal of the existing research on Asian American literature revealed that nobody up to then had consistently applied an ecocritical methodology to the ever-growing field of Asian American literature. I was so intrigued by the possibility of combining ecocriticism with my interest in Asian American

© The Author(s) 2020
B. Simal-González, *Ecocriticism and Asian American Literature*,
Literatures, Cultures, and the Environment,
https://doi.org/10.1007/978-3-030-35618-7_1

studies that, in 2007, I started work on a research project entitled "Human Ecology," which was to prove the seed for this book.[2] As Lawrence Buell argues, every literary text, consciously or unconsciously, encodes a certain type of *environmentality* (2001, 18–27, 25). There was no reason to believe that Asian American writings were an exception to the rule; however, nobody had thought to explore their "environmental unconscious" and the ecocritical potential of Asian American writings. My aim, then, was to try to fill that void, re-engaging Asian American literature with the new navigational tools of ecocriticism.

The fact that, for decades, environmental ecocritics had deftly avoided studying the fraught relationship between "nature" and "race" only made my task more complicated. Initially, I was uneasy about having to handle two concepts of such theoretical complexity and doing so in a conjoined manner. I would need to find a way to acknowledge the extralinguistic relevance of being racialized as an Asian American while circumventing the essentialist pitfalls that haunt the very construct of "race." I would then have to repeat that intellectual pirouette with the equally problematic concept of "nature." Finally, I would need to juggle both conceptual balls at the same time and try to apply those critical tools to till the seemingly arid soil of Asian American literature. Fortunately, however, environmental criticism proved an excellent plow and the Asian American texts turned out to be far more fertile than expected.

1   ECOCRITICISM: SURVEYING THE FIELD

The relatively new field of ecocriticism[3] has been variously described as "the relationship between literature and the environment conducted in a spirit of commitment to environmentalist praxis" (Buell 2005, 430) and a "branch of green studies" that examines "the relationship between human and non-human life as represented in literary texts and which theorizes about the place of literature in the struggle against environmental destruction" (Coupe 2000, 302). More recently, Louise Westling has claimed that ecocriticism seeks both to engage with contemporary literature dealing with the environment and to offer a cogent reappraisal of old literary traditions (such as the pastoral) "in light of present environmental concerns" (2014, 2). What these different definitions have in common is that they incorporate an explicit ethical and political agenda: to prevent environmental deterioration through a theoretically informed analysis of literature and culture. Nevertheless, as we will see in more detail in the last chapter, excessive emphasis on theory has also been regarded with suspicion since the beginning of ecocriticism.

Although the "green wave" of environmental activism did not acquire social visibility and prominence until 1970, with the institution of Earth Day, it can be argued that environmental concerns had already become part of the social and political agenda by the 1960s, at around the same time as other counter-cultural movements were coming into view.[4] In 1962, an "eco-book" now considered a classic of toxic discourse, Rachel Carson's *Silent Spring*, became a bestseller and a major influence for a whole generation. In the 1960s and early 1970s, we also find examples of what may be considered proto-ecocriticism or ecocriticism *avant la lettre*, including Leo Marx's *The Machine in the Garden* (1964), Raymond Williams's *The Country and the City* (1973), and Joseph Meeker's *The Comedy of Survival* (1974).

It was not until the 1990s, however, that ecocriticism became a distinct field of study within literary theory and criticism, with the creation of academic associations such as ASLE (Association for the Study of Literature and the Environment) in 1992, and its associated journal, ISLE (*Interdisciplinary Studies in Literature and the Environment*), which was first launched in 1993.[5] Two seminal books of ecocriticism also appeared in the mid-1990s: Lawrence Buell's *The Environmental Imagination* (1995) and Cheryll Glotfelty and Harold Fromm's anthology, *The Ecocriticism Reader* (1996). Although some critics continued to refer to ecocriticism "as a newly-emerging field" in the late 1990s (Love 1999; Phillips 1999; Reed 1999/2002), in the first decades of the twenty-first century ecocritical theory and practice has gained both respect and visibility.[6] At the same time, ecocriticism has had to respond to new challenges, among them the very nature of "nature," as discussed in the final chapter of this book. By 1989, Bill McKibben had already posited the advent of a postnatural (or, more accurately, "transnatural") era; a few years later, in 1994, Dana Phillips wondered whether nature was "necessary" at all, while in 2007 Timothy Morton urged critics to dispense with the concept altogether. In a more cautious appraisal, Ursula Heise has recently argued that we need to accept the fact that, in the new era of the Anthropocene, nature must be approached "as already pervasively domesticated" (2016, 158).

Precisely because of the changing and ever-expanding meanings of "nature," and while still recognizing the widespread use of nature-based ecocriticism as an umbrella word for disparate trends within the critical movement, prominent ecocritics such as Buell (2005) started to use the less common label of "environmental criticism" in the twenty-first century.[7] For Buell, this term better reflects the tendency to broaden the notion of "environment" to include not only the more or less unspoiled "nature" and wilderness of canonical nature writing, but also urban settings and degraded natural landscapes, a shift that, as we shall see, has been matched by a slow but inexorable accompanying effort to complement the traditional local perspective with a global, transnational one (Heise 2008a; Westling 2014, 6).

In the first decade of the twenty-first century, therefore, ecocritics started to envision the future of their field as moving beyond the already sanctioned nature writing or "environmental non-fiction" which had been so central in the initial "recuperative" stage of ecocriticism (Rigby 2002, 2015). If an environmental unconscious can be traced in every text, it seemed only reasonable to broaden the scope of study of literary ecocriticism. In 1996, Sven Birkerts pointed out the risk of "programmatic simplicity" of too literal a focus on traditional understandings of nature and argued instead for a "more inclusive idea of 'environment'" on the part of ecocritics (quoted in Armbruster and Wallace 2001, 3). In *The Greening of Literary Scholarship*, Steven Rosendale took up both of the objections raised by Birkerts and recognized that the "received nature-writing canon and the relatively small arsenal of critical approaches that have been applied to it have been too narrowly limited" (2002, xxvii). Similarly, in their introduction to the significantly titled *Beyond Nature Writing: Expanding the Boundaries of Ecocriticism*, Karla Armbruster and Kathleen R. Wallace contended that, for environmental criticism "to have a significant impact as a literary methodology beyond the study of nature writing," ecocritics had to prove that such an approach was more than relevant to critics working in other literary and theoretical fields and exploring texts other than nature writing (2001, 3–4); or, to use Morton's words, "the time should come when we ask of any text, 'What does this say about the environment?'" (as we have done in the realm of gender), instead of deciding in advance "which texts we will be asking" (2007, 5). Armbruster and Wallace not only argue for the need to go "beyond nature writing" but also maintain that such expansion is already taking place, albeit timidly, as the essays in their collection demonstrate by engaging an "enlarged" environment that comprises "cultivated and built landscapes, the natural elements and aspects of those landscapes, and cultural interactions with those natural elements," and by "applying ecocritical theories and methods to texts that might seem unlikely subjects because they do not foreground the natural world or wilderness" (2001, 4–5). Although Armbruster and Wallace's chosen scope was rather exceptional at the time, one could safely argue that applying an ecocritical analysis to literary texts other than nature writing has become the norm in contemporary environmental criticism.[8] Westling, writing in 2014, confirms that ecocriticism has finally "moved beyond earlier preoccupations with subjective experience of wild or rural places to increasing considerations of urban environments, collective social situations as those of oppressed minorities forced to live in polluted surroundings, postcolonial social and political realities, and global threats from pollution and climate change" (6).

In *The Future of Environmental Criticism*, Buell reviews the development of the ecocritical school from its inception and posits two main stages in environmental criticism: a first wave that exhibited a restricted understanding of "environment" and a second, more "socially oriented" wave that includes ideas of environmental justice[9] and ecofeminism.[10] First-wave practitioners of ecocriticism, consciously or otherwise, regarded nature and humankind as separate realms (2005, 21–22). By contrast, second-wave critics in the twenty-first century started "to question organicist models of conceiving both environment and environmentalism" by underscoring the inseparability of nature and culture, exploring the ways in which the "natural" and the "human-made" are inextricably mixed and interwoven with each other, and by arguing for a revised environmental ethics that includes the vexed issue of "environmental justice" (or "ecojustice") (22). Both first-wave and second-wave ecocritics have endeavored to make "neglected (sub)genres like nature writing or toxification narratives" visible, as well as proffering a much-needed interpretation "of environmental subtexts through historical and critical analyses that employ ready-to-hand analytical tools of the trade together with less familiar ones" (130). Last but not least, ecocritics have rescued from oblivion and/or reinterpreted subgenres and themes such as "pastoral, eco-apocalypticism, and environmental racism" (130).[11]

In order to explain the different ecocritical trends of the first decade of the twenty-first century—or, as Buell (2005) puts it, with the aim of "plotting internal disparities" (2005, 98)—the author describes a continuum along two axes (vertical and horizontal), which may be graphically represented as shown in Fig. 1.1, below (including, in blue type, how certain associations and movements would be placed in this "map," according to Buell).

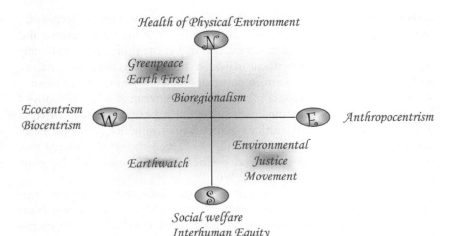

Fig. 1.1   Ecocritical trends (Buell 2005)

The shift that Buell perceived in twenty-first-century ecocriticism moves from North to South. To this initial schema, I would add a third dimension in order to incorporate the transfer of focus from local to global concerns: the "global shift" or "transnational turn," amply theorized by ecocritics like Heise (2008a, b). In the cube below, the depth line (in purple) signifies the metaphorical North-South axis described by Buell (from the northern pole of physical-natural environment to the southern pole of social environment). The line running from left to right (in green) represents the West-East axis, with the West standing for the radical ecocentrism found in deep ecology and recent theories of "bio-centered" egalitarianism[12] and the East representing the traditional anthropocentric view of humanism. The new axis I propose, running from the front to the bottom wall of the cube (in blue), signifies the shift from local (front) to global (back) concerns. The arrows used emphasize these geographic (North-South) and conceptual (Front-Back) shifts in ecocriticism (Fig. 1.2).

These two models not only reflect the situation of environmental criticism in the first decade of the twenty-first century but, as discussed in the final chapter, also highlight the future challenges for ecocriticism.[13] The two waves outlined above, however, should not be understood in "tidy" chronological terms (Buell 2005, 17) or suggest that older models and concepts are no longer valid.[14] The fact that environmental criticism has

Fig. 1.2 New ecocritical trends in perspective

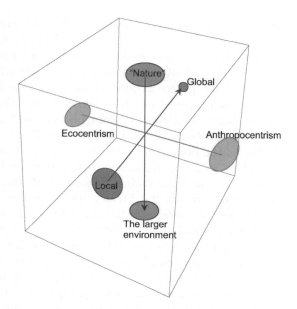

broadened its object of study and moved beyond its initial emphasis on nature writing does not mean that older concepts and discourses are in any way outworn or outdated. As the following chapters will show, both sublime and pastoral discourses continue to offer valuable strategies for writers and ecocritics alike. The versatility of the classic pastoral mode, for instance, has allowed it to mutate into the complex pastoral (Marx 1964) and the post-pastoral (Gifford 1999),[15] while new genres such as cli-fi or toxic discourse highlight the protean nature of both environmental literature and environmental criticism. In addition, new paradigms and theories have emerged—sometimes within the field of ecocriticism itself—that complement more traditional approaches.

Most prominent among the new critical schools and fields of study that coexist with(in) ecocriticism today are the different strands of posthumanist and animal studies.[16] As Axel Goodbody concisely explains in his review of European ecocritical theory, posthumanism examines "what it means to be human, and how humans relate to the natural world" (2014, 70), and is indebted to the work of philosophers such as Deleuze and Guattari, Derrida, and Agamben (Goodbody 2014, 71–72). Critical posthumanism, on the other hand, seeks to question "the discursive, institutional and material structures and processes that have presented the human as unique and bounded … above other life forms," building upon earlier critiques of humanism which highlighted "the constructedness of the normal human" (Nayar 2014, 29–30). For Pramod Nayar, the critical posthumanist paradigm that has gained ground in recent years is heir to critical humanist schools such as poststructuralist feminism and critical race studies, but also draws insights from more recent (sub)fields such as technoscience studies, monster studies, and animal studies (80–81).

Just as the boundaries between posthumanism and animal studies are often hard to define, the connections between animal studies and ecocriticism are also far from straightforward. While the objects of study of environmental criticism and literary animal studies may appear to overlap, sometimes the concerns of the two fields are at odds with each other, just as animal rights advocates sometimes clash with environmentalist activists.[17] The study of animals can arguably be regarded as a subfield of ecocriticism: nonhuman animals are, after all, a central, constitutive part of what has traditionally been considered "nature." In her overview of environmental literary criticism, Westling explains that the scope of the field ranges from "poststructural critiques of nostalgia and theoretical naiveté in nature writing" to "critical animal studies and literary animals" (2014, 2). In her contribution to Westling's volume, Sarah McFarland approaches animal studies as a central part of ecocriticism whose purpose is to "interrogate the human/animal aspects of the self/other binary and the arising consequences to subjectivity and species definitions" (2013, 153). Nevertheless, the field of animal studies is not always viewed as an epiphenomenon of ecocriticism. Certainly, the concept of animality has been the object of critical and philosophical enquiry independently of environmental criticism and, in some cases, prior to the institutionalization of ecocriticism in the late twentieth century.[18] This may explain why, in his influential "Human, All Too Human: 'Animal Studies' and the Humanities" (2009), Cary Wolfe frames animal studies within the larger discipline of cultural studies, with no reference whatsoever to the field of ecocriticism (565).[19]

Be that as it may, interest in animal studies and, more specifically, in tracking the presence and role of animality in literature has flourished in recent years.[20] One of the benefits of focusing on literary representations of nonhuman animals is to foreground the complexity of human/animal dialectics: an "entanglement" between human and nonhuman species that, according to Roman Bartosch (2016, 264), allows us to eschew both reductive "sameness"—what Heise terms the "flattening of ontologies"

(2016, 197)—and the absolute alterity that Val Plumwood (1993) theorizes as "hyperseparation." Plumwood's philosophical concerns resurface in Wolfe's work on animality, which engages in a reconfiguration of the human-animal divide. According to Wolfe, scholars working within the critical framework of animal studies do not merely approach animality as "a theme, trope, metaphor, analogy, representation, or sociological datum"; they also scrutinize and question "the knowing subject and its anthropocentric underpinnings" (2009, 567–568). Following Derrida's work on the human-animal divide (2002, 2006), Wolfe claims that both the lack of a human subjectivity prior to discourse and the vulnerability shared by human and nonhuman animals as embodied beings signal the overcoming of the humanist paradigm (2009, 571). Despite Wolfe's invocation of Derrida, it bears reminding that the theorization of embodiment owes more to the philosophical school of phenomenology (especially Maurice Merleau-Ponty's overcoming of anthropocentrism; see Westling 2014, 7; Goodbody 2014, 65–67) than to the Derridean brand of poststructuralism. The phenomenological emphasis on shared vulnerability, finitude, and embodiment, already present in Katherine Hayles's *How We Became Posthuman* (1999), has reemerged in recent work in the field of animal studies.[21] While proponents of critical posthumanism emphasize "the embedding of embodied systems in environments where the system evolves with other entities, organic and inorganic, in the environment in a mutually sustaining relationship," a vision that understands alterity "as constitutive of the human/system" (Nayar 2014, 51), scholars working in the field of animal and human-animal studies similarly highlight the shared embodiment of sentient beings.

This foregrounding of the continuum and interconnectedness between different materialities and "embodied systems," especially its emphasis on permeability (Nayar 2014, 45–46, 48), conjures up the idea of "trans-corporeality" as fleshed out by Stacy Alaimo (2008, 2010), a concept that will prove pivotal in several chapters of this book. In her critical approach, Alaimo, like other materialist ecocritics, attempts to transcend the nature/culture dichotomy that has had such negative consequences for both feminist and environmentalist discourses. In "Trans-Corporeal Feminisms and the Ethical Space of Nature" (2008), as well as in her more recent book, *Bodily Natures: Science, Environment, and the Material Self* (2010), she explains how her proposed neologism captures the interpenetration and co-constitution of the human body and its physical surroundings (2010, 2). She claims that the *trans* part of her trans-corporeal approach highlights "the movement across bodies" and, in so doing, "reveals the interchanges and interconnections between human corporeality and the more-than-human" (2008, 238) that might otherwise remain hidden. This inextricable link between all embodied beings, human or nonhuman, is what lies at the core of human-animal studies, a label that some critics have recently favored over the heretofore more common term, animal studies. The main aim of this reconceived field of human-animal studies, as Sherryl Vint notes in *Animal Alterity*, is to lay bare "the degree to which our entire philosophical tradition of subjectivity has been premised upon the separation of human from animal" (2010, 5–6).[22]

In his introduction to *Animalities: Literary and Cultural Studies Beyond the Human* (2017), Michael Lundblad tries to map out this increasingly complex field. He does so by replacing the umbrella term, "animal studies," which he considers too vague, with three labels corresponding to three distinct disciplines: "human-animal studies," "animality studies," and "posthumanism," each with their own different subfields and primary focus (2–10).[23] However, despite the apparently clear-cut nature of his classification, the overlapping between fields makes it hard to apply in all cases. For instance, according to Lundblad's taxonomy, most of the discussion in Chaps. 3 and 5 would correspond to "animality studies," which focuses on "discourses of animality in relation to human cultural politics," including the manner in which humans are constructed as nonhuman animals, and encompasses subfields such as ecofeminism and environmental justice (3, 10). Problematically, however, the biopolitical reading of animality as "bare life" in Chap. 5 corresponds more accurately to the field of posthumanism.

The solution to the increasing complexity of the field may lie not in the formulation of new labels and subfields, but in maintaining the original term, "ecocriticism." After all, contemporary environmental literary criticism continues to analyze texts "that engage environmentalist perspectives or imagine ecological catastrophe," while also unraveling "the very categories of the human and of nature" (Westling 2014, 2) and studying all the biozens on earth. While these last two critical endeavors have also been claimed by posthumanism, animal studies, and human-animal studies, the widening scope of ecocriticism both encompasses and accounts for the heterogeneity of approaches within it, as prophesied by Buell.

2 RACIALIZED OTHERS: THE ONES LEFT OUT

Environmental justice is premised on the realization that, as Jonathan Bate succinctly puts it, "ecological exploitation is always coordinate with social exploitation" (2000, 48),[24] the kind that preys on minoritized collectives such as women, impoverished and colonized peoples, or the many "racial others" that we encounter in the United States and its literature. Nevertheless, years after environmental criticism first emerged as a field of literary and cultural studies, comparatively little attention had been paid to texts written by "ethnic" minorities in America, with the obvious exception of Native American literature.[25] Aside from a few significant articles at the turn of the twenty-first century focusing on multiethnic literature (Murphy 1997, Bennett 2001), Jonathan Levin still lamented the "relative dearth of attention to ethnic American writing" among ecocritics (2002, 181), while Armbruster and Wallace wondered "why so few African American voices [were] recognized as part of nature writing and ecocriticism" (2001, 2).[26] This scarcity of work was even more conspicuous in the case of Asian American literature.[27] Arguably, then, Asian American ecocriticism is a rather recent approach, with barely a decade of work behind it. In 1992, in her groundbreaking study of environmental history, "Disorientation and Reorientation," Patricia Limerick voiced the need to address the interactions between Asian Americans and the American environment. In answer to her call, in 2007 Robert Hayashi published *Haunted by Waters*, a book that examined the ways in which Asian Americans viewed their relationship with the natural environment. Hayashi's pioneering study was a valuable effort, even though its scope remained limited in both geographic and thematic terms.[28] That same year, Hayashi published "Beyond Walden Pond: Asian American Literature and the Limits of Ecocriticism," where he again argued for the need to expand the "ecocritical canon" beyond Euro-American writings.

The lack of ecocritical attention to ethnic literature is slowly but assuredly being redressed.[29] In the last ten years, the presence of ethnoracial issues in environmental criticism has become more visible, while scholars in ethnic studies have likewise started to pay more attention to environmental matters, a trend confirmed by the special issue on ethnicity and ecocriticism produced by *MELUS Journal* in 2009, edited by Joni Adamson and Scott Slovic. Notwithstanding this "ethnic revival," the first book-length volume devoted to an ecocritical interpretation of Asian American literary texts did not appear until 2015, with the publication of

Asian American Literature and the Environment, a collection of articles edited by Lorna Fitzsimmons, Youngsuk Chae, and Bella Adams. Until then, the relatively rare contributions to the study of Asian American literature from an ecocritical perspective consisted of isolated analyses of obviously "environmentalist" novels, such as Ruth Ozeki's *My Year of Meats* and Karen Tei Yamashita's *Through the Arc of the Rain Forest.*[30]

The present book aims to offer an ecocritical reading of Asian American literature, not so much along thematic lines, as Fitzsimmons, Chae, and Adams have recently done,[31] but as an integrated appraisal of the Asian American literary tradition, from the pioneering fiction of Edith Eaton to more recent additions to the canon, such as Ozeki's *A Tale for the Time Being* (2013). However, since it is beyond the scope of a single book to cover more than a century of Asian American literature in a cogent, critical manner, I have restricted my corpus to representative narratives penned by Chinese American and Japanese American writers, even if I briefly mention other Asian American authors for comparative purposes. While not ignoring the differences between these writers, I believe that much can be gained from an ecocritical study of the Asian American literary tradition, a study that is consciously ethnic-group-specific at the same time that it remains aware of the increasingly globalized, transnational, and "transethnic" context in which it necessarily exists.

Although most of the omissions from my selection are due to the sacrifices common to any process of curation, there are some important editorial choices that bear clarification. The first difficult decision pertains to genre. With the exception of the "prelude" (Chap. 2), I have restricted my analysis to narrative texts over other genres such as poetry or drama. In electing to concentrate on prose narrative, I am conscious that I have ignored significant alternative literary forms that would have enriched my examination of the subject. By way of compensation, I have attempted to include novels and autobiographical narratives from different historical periods and diverse gender, class, and age groups. The second conspicuous limitation is the lack of an accurate reflection of the increasing complexity and internal diversity (at least in terms of national origins) of the Asian American community.[32] Once more, with the exception of the prelude, the texts covered in this book have all been sampled from Chinese American and Japanese American writers to the exclusion of writings by Korean American, Filipino American, Vietnamese American, or South Asian American authors, to name but a few. A more inclusive approach, however, would have required additional chapters or extending some of the existing ones beyond the compass of a single book. This conscious omission constitutes, in fact, an open invitation for further ecocritical research on Asian American writers other than Chinese American and Japanese American authors.

The last important choice involved in the selection of the literary corpus was the privileging of older, "canonical" texts over more recent ones. With the exception of Miyake's, Kadohata's, and Ozeki's novels, the narratives analyzed in this study all date from the twentieth century. The fact that, in most cases, these books appeared before ecocriticism itself entered the scene is, however, the very reason why I have chosen to revisit them, feeling that they need and deserve the ecocritical attention that more recent writings are receiving.[33] Moreover, by tracking the emergence and development of Asian American literature through a new critical lens, I also wish to demonstrate the presence of a certain literary "genealogy" in the texts, excavating and scrutinizing anew the most important literary milestones in the Asian American tradition: the work of the controversial literary pioneer, Edith Eaton, at the turn of the twentieth century (Chap. 3); the master narratives of cultural nationalism which attempted to "claim America" in the 1970s and 1980s (Chap. 4); "internment" narratives, dealing with the incarceration of Japanese Americans during the Second World War (Chap. 5); and the new Asian American literature that emerged from the "diasporic" or "denationalizing" impulse at the turn of the twenty-first century and coexists with earlier cultural projects (Chap. 6). As I hope these chapters will illustrate, approaching each of these writings from the perspective of environmental criticism provides unexpected insights, even (and especially) when the ecocritical reading is far from obvious and we need to dig deeper to uncover the text's environmental unconscious.

All Asian Americans, even if they are not immigrants themselves, have a history of immigration to which they are indebted. When they or their ancestors from Asia traveled to the United States, in the nineteenth or twentieth century, they became "American"—symbolically, if not always legally. Since this book is premised on the need to acknowledge the interaction of Asian Americans with their environment, it seemed fitting to start with those first-contact narratives that signaled the immigrant's initial encounter with the new materiality of the American territory, often mediated by disorienting fog. To preface this ecocritical examination and reappraisal of the major landmarks in Asian American literature, therefore, I have included what I consider to be a necessary "prelude," Chap. 2, in which I have also been able to explore a wider sample of textual fragments by a more diverse array of authors (Filipino American, Chinese American, South Asian American) than those covered in the longer chapters.

This brief prelude precedes four longer movements, in the form of full chapters, and a final coda. While the prelude follows the first Asian Americans as they entered "Nature's Nation," the first movement (Chap. 3) is devoted to the pioneer Asian American writer, Edith Eaton. Eaton first began publishing at the turn of the twentieth century, in a context of legal exclusion, Orientalist fascination, and xenophobic nativism, and even though her work often echoed the cultural conventions of the time, most notably the sentimental idiom, she showed a staunch commitment to the Chinese immigrants who had settled in North America. An ecocritical approach to Eaton's writings allows us to see that her handling of the rhetorical strategies and conventions available to her went some way toward debunking anti-Chinese sentiment among her readers. In particular, I argue that Eaton's work contributed to dismantling the racist animalization of Chinese Americans at the time, sometimes by exposing its crude fallacy and unfairness, sometimes by resorting to alternative natural metaphors.

In the fourth chapter, I move on to the 1970s and 1980s and focus on two Chinese American writers, Maxine Hong Kingston and Shawn Wong, whose works enlisted environmental strategies to further the Asian American cultural-nationalist project and, in particular, its effort to "claim America" for Asian Americans. Kingston's *China Men* and Wong's *Homebase* approach cultural nationalism differently: while *Homebase* apparently adheres to cultural-nationalist tenets in its depiction of the environment, *China Men* is rather more ambiguous in its engagement with cultural nationalism, inasmuch as the apparent nationalist agenda in the book is complicated by an emerging environmentalist subtext. My ecocritical reading of these books will pivot around two key notions: the concept of land empathy and a process of land incorporation that I refer to as "inlanding." These revised ecocritical tools emerge from the realization that some Asian American texts exceed not only the most common ecocritical methods but also the very master narratives that they have traditionally been associated with.

Probably the most famous master narrative in Japanese American literature revolves around the massive incarceration of Japanese Americans after Pearl Harbor, which has been textualized in numerous novels and autobiographies. The four books analyzed in the fifth chapter—Monica Sone's

Nisei Daughter (1953), Jeanne Wakatsuki Houston's *Farewell to Manzanar* (1973), Perry Miyake's *21st Century Manzanar* (2002), and Cynthia Kadohata's *Weedflower* (2008)—all deal with the same historical event (the "internment" experience), but they belong to different genres and were published in different historical periods. In order to tease out the similarities and differences between these internment narratives, I have combined the theoretical approaches of biopolitics and material ecocriticism. Reading the historical experience of "internment" as a literal and figurative illustration of Agamben's campo paradigm, I examine the Japanese American characters' relationship with their new environment and trace their evolution from marginal *homo sacer* to the more empowering figures of *homo faber*, *homo agricola*, and *homo ecologicus*. The formal analysis of the texts is also enhanced by the use of Alaimo's concept of trans-corporeality, which proves particularly useful in describing the change in the prisoners' attitudes toward their natural environment.

In contrast to the first chapters of this book, devoted to writers who had reached "canonical" status in Asian American studies by the end of the twentieth century, and Chap. 5, focusing on what is consensually one of the main "master narratives" of Asian America, the last full chapter of the book analyzes the work of two contemporary writers, Karen Tei Yamashita and Ruth Ozeki, who consciously depart from those traditional Asian American narratives. This chapter is an attempt to set out not only the new challenges in Asian American literature but also those faced by current environmental criticism. The external ecological crisis threatening nature at the turn of the twenty-first century was mirrored by an internal crisis in the field of ecocriticism in the form of a radical questioning of its central concept, "nature," and the widening of the canon. Ecocriticism thus witnessed several paradigm shifts, most notably the transnational and transnatural turns. The sixth chapter examines how these theoretical changes affect the way we approach a literary text as ecocritics and, conversely, how texts like Yamashita's and Ozeki's novels highlight the challenges that ecocriticism has to face. Both Yamashita and Ozeki grapple with the phenomenon of globalization and the "end of nature," and they do so, among other things, by introducing in their narratives miniature planets that float adrift or follow the characters, telling stories. By the time we reach the coda (Chap. 7), we may have learned to pierce the fog that permeates the prelude and may finally be able to glimpse the (also) "murky globes" washing on our shores.

NOTES

1. According to the WWF (World Wildlife Fund) report issued in 2003, the ecological impact of the Prestige oil spill was even worse than that of the Exxon Valdez disaster: "71,000 short tons of oil were spilled by the Prestige—60 percent more than initially estimated—contrasted with 42,769 short tons spilled by the Exxon Valdez" (https://www.worldwildlife.org/press-releases/the-prestige-one-year-later-proves-worse-than-exxon-valdez). A similar conclusion is reached in Vince's *New Scientist* article: https://www.newscientist.com/article/dn4100-prestige-oil-spill-far-worse-than-thought.

2. As tends to be the case, life interfered with the initial project and delayed its completion. In the meantime, a limited number of valuable ecocritical publications in the field of Asian American studies have appeared, as will be explained later in more detail.

3. For an earlier, shorter version of the following discussion, see Simal (2010).

4. In "Give Earth a Chance," Adam Rome argues that the emergence of environmentalism owes much to the interaction of three phenomena taking place in the 1960s: "the revitalization of liberalism, the growing discontent of middle-class women, and the explosion of student radicalism and countercultural protest" (2003, 527).

5. The past decade has witnessed the proliferation of ASLE-type organizations around the world, such as the European Association for the Study of Literature, Culture and the Environment (EASLCE, 2007), the Association for Literature, Environment and Culture in Canada (ALECC, 2007), and so on, while branches of ASLE have appeared in countries such as Japan (1994), the UK (1999), Korea (2001), Australia and New Zealand (2005), India (ASLE-India, renamed OSLE in 2006), and Taiwan (2008). See Westling (2014, 241–246) for a full chronology of the discipline.

6. Compare the somewhat bleak panorama drawn by Cheryll Glotfelty (1996) at the beginning of her "Introduction" to *The Ecocriticism Reader* (especially the disparity she observes between the attention given by the media to the environmental crisis and the total neglect on the part of literary critics; xvi), with the more optimistic, though by no means exultant view of ecofeminism offered two years later by Greta Gaard and Patrick D. Murphy (1998), in *Ecofeminist Literary Criticism*: "In the 1990s ecofeminism is finally making itself felt in literary studies" (5). Speaking from a midposition, in 2005 Lawrence Buell contended that, although the debate around environmental issues, most poignantly our (humans') relationship to Nature, is as old as the *Book of Genesis*, ecocriticism was still struggling for visibility and recognition in academic circles (1–2).

7. Buell (2005 12, 21, 82, 90–92). Phillips likewise contends that the term ecocriticism is "just a troublesome as it is helpful" (2003, viii), and that

nature continues to be "one of philosophy's least precise and most contested terms" (32). On the other hand, we cannot forget that, for ecocritics such as Glotfelty, the term "environment" still has pejorative, "anthropocentric and dualistic" connotations, whereas the prefix "eco-" suggests positive images of "interdependent communities, integrated systems" (1996, xx). All in all, I do agree with Buell's reasons for favoring the term "environmental criticism," even though ecocriticism continues to be the most widespread label.

8. Recent evidence of such a trend can be found in the special issue that *American Literature* devoted to environmental criticism in 2012. As the editors of the volume put it in their preface, the contributions to this issue try to steer "away from the place-based, policy-focused, and phenomenological preoccupations of older forms of ecocriticism toward an engagement with murkier aspects of our condition," such as "the significance of new identities and communities—often involuntary—that arise from environmental crisis; and the complex new forms of affect that accompany the recategorization of the planet" (Allewaert and Ziser 2012, 235).

9. In 2002, *The Environmental Justice Reader*, edited by Joni Adamson, Mei Mei Evans, and Rachel Stein, definitively launched the question of ecojustice. In their groundbreaking introduction, the editors defined environmental justice as "the right of all people to share equally in the benefits bestowed by a healthy environment," while the environment would encompass all those "places in which we live, work, play, and worship" (1).

10. As Noël Sturgeon succinctly puts it in *Ecofeminist Natures* (1997), the ecofeminist "movement … articulates the theory that the ideologies that authorize injustices based on gender, race and class are related to the ideologies that sanction the exploitation and degradation of the environment" (23). Annette Kolodny's *The Lay of the Land* (1975) can be considered the foundational text of ecofeminism. Among the key books that advanced ecofeminist scholarship in the 1990s, it is worth mentioning Carolyn Merchant's *The Death of Nature* (1990), Val Plumwood's *Feminism and the Mastery of Nature* (1993), and, in the particular field of American literature, Louise Westling's *The Green Breast of the New World* (1996).

11. Some ecocritics have recently posited a "third wave" of ecocriticism (see Wald 2016, 13), but most of the concerns they highlight as new, such as the focus on ecofeminist and ethnoracial issues, were already prominent in the second wave described by Buell. However, it is too early to decide whether this second wave, which still emerged at the turn of the twenty-first century, may derive into a new phase in the field of ecocriticism in the near future.

12. "Ecocentrism" views our planet as an interconnected community with no boundaries between sentient and non-sentient beings, humans and nonhu-

mans, since we all depend on one another. Buell defines ecocentrism as the belief "that the interest of the ecosphere must override that of the interest of individual species," in contradistinction with anthropocentrism (2005, 137). A similar, but more recent concept is bio-centered egalitarianism, as theorized by materialist ecocritic Rosi Braidotti (2006). This "trans-species" egalitarianism (41) is a hallmark of much of animal studies, as will be shown later.

13. Westling's 2014 overview of the present state of the field highlights very similar trends (6–7).

14. It cannot be sufficiently emphasized that, despite the precise contours of the cube metaphor, there are no clear-cut dichotomies, but lines signifying a mesh or continuum, a fluid, crisscrossing space in which the different positions interpenetrate each other.

15. Indeed, the pastoral's flexible or "commodious" nature may explain why it is often "defined as a mode rather than as a genre": not only does it "assume more than one form" but it can also "serve more than one master" (Phillips 2003, 17).

16. I thank the anonymous reader for pointing out the importance of both posthumanist theory and the "animal turn" in recent years.

17. For an insightful overview of the history of these tensions, see Heise (2016, 128–140). According to Heise, the two groups' radically divergent approaches to the human-animal relationship may be accurately described as paradigmatic, in the case of animal rights activism, and syntagmatic, in the case of environmentalist activism (143–144).

18. See classical philosophers' work (most notably Aristotle's) on the human-animal divide. For more recent philosophical examinations of animality, see Deleuze and Guattari's discussion in *A Thousand Plateaus* (1987), Giorgio Agamben's *The Open: Man and Animal* (2002), Jacques Derrida's *The Animal That Therefore I Am* (2008), Donna Haraway's *When Species Meet* (2007), and Mathew Calarco's *Zoographies* (2008).

19. Interestingly, recent animal studies scholars rarely assert their ecocritical genealogy either, preferring the philosophical tradition outlined in the previous note. Wolfe, for instance, claims that Derrida's 2002 essay "The Animal That Therefore I Am (More to Follow)" and the homonymous book translated in 2008 constituted "the single most important event in the brief history of animal studies" (2009, 570).

20. Some of the most important contributions to literary animal studies include (in chronological order) Margot Norris's *Beasts of the Modern Imagination* (1985), Steve Baker's *The Postmodern Animal* (2000) and *Picturing the Beast* (2001), and Cary Wolfe's *Animal Rites* (2003a) and *Zoontologies* (2003b). For an overview of the essential bibliography in critical animal studies, see Wolfe's "Human, All Too Human" (2009). More

recent publications, not covered in Wolfe's essay, include Sherryl Vint's *Animal Alterity* (2010), Anat Pick's *Creaturely Poetics* (2011), Ursula Heise's *Imagining Extinction* (2016), Mario Ortiz-Robles's *Literature and Animal Studies* (2016), and Michael Lundblad's *Animalities* (2017). To this list we need to add the latest contributions to the Palgrave Series "Studies in Animals and Literature," such as *Creatural Fictions*, edited by David Herman in 2016, and *Beyond the Human-Animal Divide*, edited by Dominik Ohrem and Roman Bartosch in 2017. For a discussion of the "creaturely discourse," see Herman (2016, 3–6).

21. See Ralph Acampora's concept of *symphysis* (quoted in Vint 2010, 15) and Erica Fudge's privileging of the common embodied experience of human and nonhuman animals in *Pets* (2008).

22. Vint also contends that science fiction is a privileged site for the exploration of how alterity is constructed (2010, 1), including rethinking the boundary between the human and animal categories (8). Additionally, science fiction allows us "to put ourselves in the place of the animal other and experience the world from an estranged point of view," thus fostering "sympathetic imagination" in human readers (15).

23. The fact that there is still no critical consensus concerning the boundaries and priorities of animal studies and posthumanism can be seen if we compare the positions of three recently published books. In Heise's *Imagining Extinction* (2016), the critic claims that both advocates of animal rights and "the posthumanist variant in recent philosophy" attempt "to establish animals as members of the human social, political, and legal community" (148). In *Animalities* (2017), Lundblad argues that posthumanism seems less "engaged in situated and historicized political advocacy and activism" than animality studies and human-animal studies (11). According to Lundblad, the field of human-animal studies encompasses both concern for the welfare of animals and the exploration of inter-species relationships (3); for Nayar (2014), however, these two objects of study correspond to two different disciplines (93).

24. This realization is a hallmark of postcolonial and environmental studies, as seen in the pioneering efforts of Joan Martínez Alier and Ramachandra Guha in the twentieth century, and in more recent and equally influential books by Rob Nixon, Graham Huggan, and Helen Tiffin. As we shall see in Chap. 3, some of the insights gained from the overview of animal studies prove rather helpful in our discussion of environmental justice, given that "discourses of animality have long been used to oppress various human groups" (Lundblad 2017, 8).

25. This is associated with the image of the "Ecological Indian" (Krech 1999); see Chap. 4 for a discussion of this trope.

26. As early as 2009, Julie Sze argued that this was a plausible project, since "ethnic literature can be explored with an ecocritical lens, prioritizing shared themes in ethnic and environmental texts such as loss, mourning, and death, and also renewal in and through the land" (2009, 200). I find this to be valuable advice, even though I do not cover all of the "shared themes" mentioned in her article. See also contributions by Dixon, Dodd, Murphy, and Parra to the *PMLA* "Forum on Literatures of the Environment" (Arnold et al. 1999).

27. The first attempts to turn the focus onto ethnic traditions other than the Euro-American (such as the MELUS issue guest-edited by Slovic and Adamson in 2009 or Slovic's anthology *Getting Over the Color Green*) contain only a few references to Asian American literature. Similarly, while in her 2002 survey of ecocriticism Kate Rigby noted a tendency toward broadening the spectrum of enquiry, she too failed to mention Asian Americans in her list (164).

28. The spatial restrictions of the work derive from the author's choice to concentrate on just one Western state (Idaho). As to its disciplinary scope, even though the book does include some poems and references to Sone's *Nisei Daughter*, it is far more interested in environmental history than in literary analysis. Nevertheless, Hayashi's *Haunted by Waters* constitutes a highly readable journey across the landscapes of Idaho: a fishing excursion in which the author discusses the role of race in the environmental history of the Gem State. In his study-cum-travel-diary, Hayashi provides important insights into the limits of agrarianism, the construction of the ethnic/racial other (Shoshone, Chinese, Japanese, etc.) as weeds or vermin, and the (mis)use of the pioneer discourse during the internment years, with special attention paid to the Minidoka camp in south Idaho.

29. A number of book-length publications focusing on "ethnic" American literature have appeared in the last few years. John Gamber's *Positive Pollutions and Cultural Toxins* (2012), Jennifer Ladino's *Reclaiming Nostalgia* (2012), Sarah Wald's *The Nature of California* (2016), and Molly Wallace's *Risk Criticism* (2016) discuss a variety of texts from different ethnoracial backgrounds, including Asian American. Other scholars have chosen to focus on African Americans or continue their exploration of Native American culture: Kimberly Smith's *African American Environmental Thought* (2007), Paul Outka's *Race and Nature from Transcendentalism to the Harlem Renaissance* (2008), Ian Finseth's *Shades of Nature: Visions of Place in the Literature of American Slavery* (2009), Kimberly Ruffin's *Black on Earth: African American Ecoliterary Traditions* (2010), and Joni Adamson's *American Indian Literature, Environmental Justice, and Ecocriticism* (2001), Lindsey Smith's *Indians, Environment, and Identity on the Borders*

of American Literature (2008), or Lee Schweninger's *Listening to the Land* (2008). Just as this manuscript was being sent to press, a collection of articles on Yamashita's work, edited by A. Robert Lee, came out (I thank the editorial reviewers that alerted me to this).

30. Indeed, in his introduction to the 2015 compilation, John Gamber wondered why it had taken "so long" for an ecocritical reading of Asian American literature to appear, given that Asian American texts are "replete with narratives that focus on ecological connections" (1). In contrast to the relative scarcity of book-length publications devoted to Asian American literature, East Asian literatures and cultures have garnered more attention in the field of environmental criticism and have recently been the subject of analyses from an ecocritical perspective in monographic studies such as Thornber's *Ecoambiguity: Environmental Crises and East Asian Literatures* (2012) and Estok and Kim's *East Asian Ecocriticisms* (2013).

31. Fitzsimmons, Chae, and Adams's groundbreaking compilation of articles is organized around three thematic foci: The Environment and Labor, The Environment and Violence, and The Environment and Philosophy. The ten chapters in the volume cover Asian American authors from different periods and ethnicities, thus reflecting the ever-growing diversity of Asian American literature. It is worth noting that only a few of the texts studied in their collection are also discussed in this book: Kingston's *China Men* (also the subject of a previous article of mine: Simal 2013) and Houston's internment memoir, *Farewell to Manzanar*.

32. This conflation of people with different national origins into a Pan-Asian American community, as Sau-ling Wong (1993) rightly points out, should be "*voluntarily adopted* and highly *context-sensitive* in order to work" (6; italics in the original). In other words, the nature of this large Pan-Asian alliance is temporary and its function is strategic, "merely to provide an instrument for political mobilization under chosen circumstances" (6). On the necessary acknowledgment of internal heterogeneity, see also Lowe (1996).

33. Several books published by Asian American writers in the twenty-first century invite an ecocritical reading by explicitly including environmental concerns: David Mas Masumoto's non-fiction, Amitav Ghosh's *The Hungry Tide* (2004), or Don Lee's *Wrack and Ruin* (2008). These last two novels, penned by a South Asian and a Korean American, respectively, highlight the increasing diversity of Asian Americans. However, as both texts have already been analyzed from an ecocritical perspective, I have chosen not to include them in my study. Bhira Backhaus's novel *Under the Lemon Trees* (2009) would also have been an excellent addition to my corpus. I thank the anonymous readers for these suggestions.

REFERENCES

Adamson, Joni. 2001. *American Indian Literature, Environmental Justice, and Ecocriticism: The Middle Place*. Tucson: University of Arizona Press.

Adamson, Joni, and Scott Slovic, eds. 2009. Ethnicity and Ecocriticism, Special Issue, *MELUS* 34 (2): 5–24.

Adamson, Joni, Mei Mei Evans, and Rachel Stein, eds. 2002. *The Environmental Justice Reader: Politics, Poetics, and Pedagogy*. Tucson: The University of Arizona Press.

Agamben, Giorgio. (2002) 2004. *The Open: Man and Animal*. Stanford: Stanford University Press.

Alaimo, Stacy. 2008. Trans-Corporeal Feminisms and the Ethical Space of Nature. In *Material Feminisms*, ed. Stacy Alaimo and Susan Hekman, 237–263. Bloomington: Indiana University Press.

———. 2010. *Bodily Natures: Science, Environment, and the Material Self*. Bloomington: Indiana University Press.

Allewaert, M., and M. Ziser. 2012. Preface: Under Water. *American Literature* 84 (2): 233–241.

Armbruster, Karla, and Kathleen R. Wallace, eds. 2001. *Beyond Nature Writing: Expanding the Boundaries of Ecocriticism*. Charlottesville: University Press of Virginia.

Arnold, Jean, et al. 1999. Forum on Literatures of the Environment. *PMLA* 114 (5): 1089–1104. https://doi.org/10.2307/463468.

Baker, Steve. 2000. *The Postmodern Animal*. London: Reaktion.

———. 2001. *Picturing the Beast: Animals, Identity, and Representation*. Urbana: University of Illinois Press.

Bartosch, Roman. 2016. Ghostly Presences: Tracing the Animal in Julia Leigh's *The Hunter*. In *Creatural Fictions*, ed. David Herman, 259–275. New York: Palgrave Macmillan.

Bate, Jonathan. 2000. *The Song of the Earth*. Cambridge: Harvard University Press.

Bennett, Michael. 2001. Anti-Pastoralism, Frederick Douglass, and the Nature of Slavery. In *Beyond Nature Writing: Expanding the Boundaries of Ecocriticism*, ed. Karla Armbruster and Kathleen R. Wallace, 195–209. Charlottesville: University Press of Virginia.

Braidotti, Rosi. 2006. *Transpositions: On Nomadic Ethics*. Cambridge: Polity.

Buell, Lawrence. 1995. *The Environmental Imagination: Thoreau, Nature Writing, and the Formation of American Culture*. Cambridge: Harvard University Press.

———. 2001. *Writing for an Endangered World: Literature, Culture, and Environment in the United States and Beyond*. Cambridge: Harvard University Press.

———. 2005. *The Future of Environmental Criticism: Environmental Crisis and Literary Imagination*. Oxford: Blackwell.

Calarco, Mathew. 2008. *Zoographies: The Question of the Animal from Heidegger to Derrida*. New York: Columbia University Press.

Coupe, Laurence. 2000. *The Green Studies Reader: From Romanticism to Ecocriticism*. New York: Routledge.

Deleuze, Gilles, and Félix Guattari's. 1987. *A Thousand Plateaus: Capitalism and Schizophrenia*. Minneapolis: University of Minnesota Press.

Derrida, Jacques. 2002. The Animal That Therefore I Am (More to Follow). *Critical Inquiry* 28 (2): 369–418.

———. (2006) 2008. *The Animal That Therefore I Am*. New York: Fordham University Press.

Estok, Simon C., and Won-Chung Kim, eds. 2013. *East Asian Ecocriticisms: A Critical Reader*. New York: Palgrave Macmillan.

Finseth, Ian Frederick. 2009. *Shades of Green: Visions of Nature in the Literature of American Slavery, 1770–1860*. Athens: University of Georgia Press.

Fitzsimmons, Lorna, Youngsuk Chae, and Bella Adams, eds. 2015. *Asian American Literature and the Environment*. New York: Routledge.

Fudge, Erika. (2008) 2014. *Pets: The Art of Living*. New York: Routledge.

Gaard, Greta, and Patrick Murphy. 1998. Introduction. In *Ecofeminist Literary Criticism: Theory, Interpretation, Pedagogy*, eds. Greta Gaard and Patrick Murphy, 1–13. Urbana and Chicago: University of Illinois Press.

Gamber, John Blair. 2012. *Positive Pollutions and Cultural Toxins: Waste and Contamination in Contemporary U.S. Ethnic Literatures*. Lincoln: University of Nebraska Press.

———. 2015. Introduction: Ecocriticism and Asian American Literature. In *Asian American Literature and the Environment*, ed. Lorna Fitzsimmons, Youngsuk Chae, and Bella Adams, 1–9. New York: Routledge.

Gifford, Terry. 1999. *Pastoral*. London: Routledge.

Glotfelty, Cheryll. 1996. Introduction. In *The Ecocriticism Reader: Landmarks in Literary Ecology*, ed. Cheryll Glotfelty and Harold Fromm, xv–xxxvii. Athens: University of Georgia Press.

Goodbody, Axel. 2014. Ecocritical Theory: Romantic Roots and Impulses from Twentieth-Century European Thinkers. In *Cambridge Companion to Literature and the Environment*, ed. Louise Westling, 61–74. Cambridge: Cambridge University Press.

Guha, Ramachandra. 1989. Radical American Environmentalism and Wilderness Preservation: A Third World Critique. *Environmental Ethics* 11: 71–83.

Haraway, Donna J. 2007. *When Species Meet*. Minneapolis: University of Minnesota Press.

Hayashi, Robert T. 2007a. Beyond Walden Pond: Asian American Literature and the Limits of Ecocriticism. In *Coming into Contact: Explorations in Ecocritical Theory and Practice*, ed. Annie Merrill Ingram, Ian Marshall, Daniel J. Philippon, and Adam W. Sweeting, 58–75. Athens: University of Georgia Press.

———. 2007b. *Haunted by Waters: A Journey Through Race and Place in the American West*. Iowa City: University of Iowa Press.

Hayles, Katherine. 1999. *How We Became Posthuman: Virtual Bodies in Cybernetics, Literature and Informatics*. Chicago: University of Chicago Press.

Heise, Ursula K. 2008a. Ecocriticism and the Transnational Turn in American Studies. *American Literary History* 20 (1–2): 381–404.

———. 2008b. *Sense of Place and Sense of Planet: The Environmental Imagination of the Global*. New York: Oxford University Press.

———. 2016. *Imagining Extinction: The Cultural Meanings of Endangered Species*. Chicago: University of Chicago Press.

Helphand, Kenneth. 1997. Defiant Gardens. *The Journal of Garden History* 17 (2): 101–121. https://doi.org/10.1080/01445170.1997.10412542.

Herman, David, ed. 2016. *Creatural Fictions*. New York: Palgrave Macmillan.

Huggan, Graham, and Helen Tiffin. 2010. *Postcolonial Eco-Criticism: Literature, Animals, Environment*. London: Routledge.

Kolodny, Annette. 1975. *The Lay of the Land: Metaphor as History and Experience in American Life and Letters*. Chapel Hill: University of North Carolina Press.

Krech, Shepard, III. 1999. *The Ecological Indian: Myth and History*. New York: Norton.

Ladino, Jennifer K. 2012. *Reclaiming Nostalgia: Longing for Nature in American Literature*. Charlottesville: University of Virginia Press.

Lee, A. Robert, ed. 2018. *Karen Tei Yamashita: Fictions of Magic and Memory*. Honolulu: University of Hawai'i Press.

Levin, Jonathan. 2002. Beyond Nature: Recent Work in Ecocriticism. *Contemporary Literature* 43 (1): 171–186.

Love, Glen A. 1999. Ecocriticism and Science Consilience? *New Literary History* 30 (3): 561–576.

Lowe, Lisa. 1996. *Immigrant Acts: On Asian American Cultural Politics. Durham*. Duke University Press.

Lundblad, Michael. 2017. *Animalities: Literary and Cultural Studies beyond the Human*. Edinburgh: Edinburgh University Press.

Martínez Alier, Joan. 1992. *De la Economía ecológica al ecologismo popular*. Barcelona: Icaria.

———. 2005. *El ecologismo de los pobres: conflictos ambientales y lenguajes de valoración*. Barcelona: Icaria.

Marx, Leo. 1964. *The Machine in the Garden: Technology and the Pastoral Ideal in America*. New York: Oxford University Press.

McFarland, Sarah E. 2013. Animal Studies, Literary Animals, and Yann Martel's *Life of Pi*. In *The Cambridge Companion to Literature and the Environment*, ed. Louise Westling, 152–166. Cambridge: Cambridge University Press.

McKibben, Bill. (1989) 2006. *The End of Nature*. New York: Random House.

Meeker, Joseph. (1974) 1997. *The Comedy of Survival: Literary Ecology and the Play Ethic*. Tucson: University of Arizona Press.

Merchant, Carolyn. 1990. *The Death of Nature: Women, Ecology, and the Scientific Revolution.* San Francisco: Harper.

Morton, Timothy. 2007. *Ecology Without Nature: Rethinking Environmental Aesthetics.* Cambridge: Harvard University Press.

Murphy, Patrick D. 1997. Commodification, Resistance, Inhabitation, and Identity in the Novels of Linda Hogan, Edna Escamill, and Karen Tei Yamashita. *Phoebe: Journal of Feminist Scholarship, Theory and Aesthetics* 9 (1): 1–10.

Nayar, Pramod K. 2014. *Posthumanism.* Cambridge: Polity.

Nixon, Rob. 2011. *Slow Violence and the Environmentalism of the Poor.* Cambridge: Harvard University Press.

Norris, Margot. 1985. *Beasts of the Modern Imagination: Darwin, Nietzsche, Kafka, Ernst, and Lawrence.* Baltimore: Johns Hopkins University Press.

Ohrem, Dominik, and Roman Bartosch, eds. 2017. *Beyond the Human-Animal Divide: Creaturely Lives in Literature and Culture.* New York: Palgrave Macmillan.

Ortiz-Robles, Mario. 2016. *Literature and Animal Studies.* New York: Routledge.

Outka, Paul. 2008. *Race and Nature from Transcendentalism to the Harlem Renaissance.* New York: Palgrave Macmillan.

Phillips, Dana. 1994. Is Nature Necessary? In *The Ecocriticism Reader: Landmarks in Literary Ecology*, ed. Cheryll Glotfelty and Harold Fromm, 204–222. Athens: University of Georgia Press.

———. 1999. Ecocriticism, Literary Theory, and the Truth of Ecology. *New Literary History* 30 (3): 577–602.

———. 2003. *The Truth of Ecology: Nature, Culture, and Literature in America.* New York: Oxford University Press.

Pick, Anat. 2011. *Creaturely Poetics. Animality and Vulnerability in Literature and Film.* New York: Columbia University Press.

Plumwood, Val. 1993. *Feminism and the Mastery of Nature.* London: Routledge.

Reed, T.V. (1999) 2002. Toward an Environmental Justice Ecocriticism. In *The Environmental Justice Reader: Politics, Poetics, and Pedagogy*, ed. Joni Adamson, Mei Mei Evans, and Rachel Stein, 145–162. Tucson: The University of Arizona Press.

Rigby, Kate. 2002. Ecocriticism. In *Introducing Criticism at the Twenty-First Century*, ed. Julian Wolfreys, 1st ed., 151–178. Edinburgh: Edinburgh University Press.

———. 2015. Ecocriticism. In *Introducing Criticism at the Twenty-First Century*, ed. Julian Wolfreys, 2nd ed., 122–154. Edinburgh: Edinburgh University Press.

Rome, Adam. 2003. 'Give Earth a Chance': The Environmental Movement and the Sixties. *Journal of American History* 90 (2): 525–554.

Rosendale, Steven. 2002. *The Greening of Literary Scholarship: Literature, Theory and the Environment.* Iowa City: University of Iowa Press.

Ruffin, Kimberly N. 2010. *Black on Earth: African American Ecoliterary Traditions.* Athens: University of Georgia Press.

Schweninger, Lee. 2008. *Listening to the Land: Native American Literary Responses to the Landscape.* Athens: University of Georgia Press.

Simal, Begoña. 2010. The Junkyard in the Jungle': Transnational, Transnatural Nature in Karen Tei Yamashita's *Through the Arc of the Rain Forest*. *JTAS: Journal of Transnational American Studies* 2 (1): 1–25.

Slovic, Scott, ed. 2001. *Getting over the Color Green: Contemporary Environmental Literature of the Southwest*. Tucson: University of Arizona Press.

Smith, Kimberly K. 2007. *African American Environmental Thought: Foundations*. Lawrence: University Press of Kansas.

Smith, Lindsey C. 2008. *Indians, Environment, and Identity on the Borders of American Literature: From Faulkner and Morrison to Walker and Silko*. New York: Palgrave Macmillan.

Sturgeon, Nöel. 1997. *Ecofeminist Natures: Race, Gender, Feminist Theory and Political Action*. New York: Routledge.

Sze, Julie. 2009. Review of Haunted by Waters: *A Journey through Race and Place in the American West/I Call to Remembrance: Toyo Suyemoto's Years of Internment/ Letters to the Valley: A Harvest of Memories*. *MELUS* 34 (2): 199–202. https://search.proquest.com/docview/203672081?accountid=17197

Thornber, Karen Laura. 2012. *Ecoambiguity: Environmental Crises and East Asian Literatures*. Ann Arbor: University of Michigan Press.

Vince, Gaia. 2003. Prestige Oil Spill Far Worse Than Thought. *New Scientist, August* 27: 2003. www.newscientist.com/article/dn4100-prestige-oil-spill-far-worse-than-thought.

Vint, Sherryl. 2010. *Animal Alterity: Science Fiction and the Question of the Animal*. Liverpool: Liverpool University Press.

Wald, Sarah. 2016. *The Nature of California: Race, Citizenship, and Farming since the Dust Bowl*. Seattle: University of Washington Press.

Wallace, Molly. 2016. *Risk Criticism: Precautionary Reading in an Age of Environmental Uncertainty*. Ann Arbor: University of Michigan Press.

Westling, Louise. 1996. *The Green Breast of the New World: Landscape, Gender and American Fiction*. Athens: Georgia University Press.

———, ed. 2014. *The Cambridge Companion to Literature and the Environment*. Cambridge: Cambridge University Press.

Williams, Raymond. 1973. *The Country and the City*. Oxford: Oxford University Press.

Wolfe, Cary. 2003a. *Animal Rites: American Culture, the Discourse of Species, and Posthumanist Theory*. Chicago: University of Chicago Press.

———, ed. 2003b. *Zoontologies: The Question of the Animal*. Minneapolis: University of Minnesota Press.

———, ed. 2009. Human, All Too Human: 'Animal Studies' and the Humanities. *PMLA* 124 (2): 564–575. JSTOR. http://www.jstor.org/stable/25614299.

Wong, Sau-Ling Cynthia. 1993. *Reading Asian American Literature: From Necessity to Extravagance*. Princeton: Princeton University Press.

World Wildlife Fund. 2003. The Prestige: One Year Later Proves Worse than Exxon Valdez. November 07, 2003. https://www.worldwildlife.org/press-releases/the-prestige-one-year-later-proves-worse-than-exxon-valdez

CHAPTER 2

Prelude: Entering "Nature's Nation"

No nation is without nature.[1] However, as has been the case in many other fields, the United States has been endowed with an apparent exceptionalism in this terrain (Madsen 1998). Self-defined as "Nature's Nation,"[2] America purportedly acquired its quintessential character by its prolonged negotiation of wilderness and "civilization," as put forward in Frederick Jackson Turner's polemical frontier thesis.[3] From the vantage point of Europeans, the New World—comprising all the Americas—was indeed that: new and pristine.[4] Needless to say, the American Dream was likewise constructed around this myth of an "untouched," exuberant natural world that welcomed people from all around the globe (at the same time that it erased the very presence of indigenous inhabitants). However, this imagined—and mostly imaginary—paradise was soon perceived to be threatened by progress and industrialization, what Leo Marx brilliantly captured in the concise phrase noting the intrusion of "the machine in the garden" (1964). While both natural icons, like the bold eagle and the Yosemite Valley,[5] and artificial or human-made ones, like the Statue of Liberty or the NY skyline, remain powerful symbols of the United States, it can be argued that, after the fall of the World Trade Center towers, the now "fragile" NY skyline has been eclipsed by the enduring grandeur of Yosemite Valley. The latter seems to be less exposed to acts of massive destruction, even though this durability may likewise be illusory.

© The Author(s) 2020
B. Simal-González, *Ecocriticism and Asian American Literature,*
Literatures, Cultures, and the Environment,
https://doi.org/10.1007/978-3-030-35618-7_2

It is in this context that we need to place what we can call, for lack of a better word, "neo-first-contact narratives." While the original first-contact narratives in American literature had colonial connotations and tended to conjure up the "meeting" of two civilizations that had theretofore ignored each other, as in the Pocahontas-Smith model, later first-contact narratives comprise all manner of encounters with the new people and the new land, most notably the immigrants' arrival, but also the traveler's and the exile's first contacts with the United States. Probably one of the most striking images of such an encounter can be found in Wendy Law-Yone's novel *The Coffin Tree*:

> Even when times were hard, the life we left behind had run along a groove cut by tradition, familiarity, and habit. But arriving in New York, my brother and I fell out of that groove, and finding our footing was nearly as awkward as the astronauts' first steps in the atmosphere of the moon.
>
> We landed in America three months after they landed on the moon and watched the event on a giant television screen that hung above the maze of cosmetics and costume jewelry in a Fifth Avenue store. It was our first American department store. (44)

Not all the arrival scenes resort to hyperbolic images, like the moon-landing metaphor employed by Law-Yone, but they do resort to tropes that express longing, disorientation, and also, in a few cases, "the shock of arrival" (Alexander 1996). Although arrival scenes abound in Asian American immigration stories, I am particularly interested in those first-contact narratives that depict the materiality of the new country, its environment. Therefore, in this brief prelude, I will follow the Asian American characters as they arrive in America and experience their first contact with the new continent.[6] In these arrival scenes, America is first experienced through "natural" elements, through "artificial" constructions, or, more often than not, through a combination of both.

Gam Saan, the Gold Mountain, is a physical site, corresponding to Western North America, a region where the discovery of gold in the mid- and late nineteenth century attracted countless immigrants. However, Gam Saan is also a mental site in the communal imagination of nineteenth and twentieth century Chinese people, a "Cantonese invention," as we shall see in more detail in later chapters. The first Chinese American texts to talk about the Gold Mountain and focus on the natural environment encountered upon arrival were the poems written by immigrants on the walls of the Angel Island Station. Angel Island operated as an Immigration Station from 1910 to 1940 and the detainees, mostly Chinese, could spend from a couple of weeks to a couple of years waiting for inspection, interrogation, and, when denied entry, court appeals.[7] Many of the immigrants, especially those whose cases proved problematic and had to spend many months in this prison-like station, vented their anger and frustration in short poems, mostly anonymous, written or carved in the very walls that curtailed their freedom. Many of these wall poems were collected, translated, and published in *Island*, thanks to Him Mark Lai, Genny Lim, and Judy Yung in 1980. Around one-fifth of the bilingual poems selected by Lai, Lim, and Yung resort to the natural environment in metaphorical and literal ways.[8] In Angel Island poetry, as Patricia Nelson Limerick puts it, "one finds ample evidence that a people who might initially seem inarticulate to the historian were charged with intense and profound responses to the American landscape," which they frequently used "as an analogy for the experience of being stranded and kept waiting" (1992, 1029).

Therefore, the Angel Island poets did not just linger on their physical space of confinement; rather, many of them looked outside the window and included the natural environment that the inmates saw. In most cases, however, nature was used as an objective correlative, since these poems projected onto the environment the poet's own melancholic or frustrated mood. Such a pessimistic view of their surroundings contrasts with the more optimistic attitude found in a later text, Carlos Bulosan's *America Is in the Heart* (1946). We cannot forget that during the 1920s and 1930s, while Chinese immigrants were still being processed and held at Angel Island, Filipinos like Bulosan could migrate freely to the United States. Therefore, the different legal situation of the immigrants of these two Asian countries partly accounts for the disparity in tone. The other important reason is that the narrative strategy in *America Is in the Heart* is rather different. While this novel is also highly autobiographical, it is also a *roman a thèse* that intends to claim America as (adoptive) home, thus contributing to the cultural-nationalist project even before the emergence of the Asian American movement.[9] In juxtaposing the initial American Dream with the subsequent American nightmare, *America Is in the Heart* proves especially illuminating, and the experience of Asian immigrants' arrival in America is no exception. After a detailed description of the hardships of the sea voyage, Carlos devotes merely a paragraph to the moment of arrival, but it paints such a hopeful picture that it is worth quoting at length:

We arrived in Seattle on a June day. My first sight of the approaching land was an *exhilarating* experience. Everything seemed native and *promising* to me. It was *like coming home* after a long voyage, although as yet I had no *home* in this city. Everything seemed familiar and kind—the white faces of the buildings melting in the soft afternoon sun, the gray contours of the surrounding valleys that seemed to vanish in the last periphery of light. With a sudden surge of joy, I knew that I must find a *home* in this new land. (99; emphasis added)

Whatever the following chapters have in store for young Carlos, this scene apparently fits the convention of the American Dream immigration story. The description aptly conveys a young boy's wish for a better future, which the text underscores with positive adjectives and similes: *exhilarating*, *promising*, *like coming home*, *familiar and kind*, and so on. If we focus on the material, environmental aspects of the first contact, we notice a combination of the natural and the artificial: both the American landscape (*valleys*) and human constructions (*buildings*) seem to greet the Filipino immigrants. Intriguingly, both types of elements appear blurred by the declining light: "the white faces[10] of the buildings *melting* in the soft afternoon sun, the gray contours of the surrounding valleys that seemed to *vanish* in the last periphery of light" (99). While this disappearance conjures up a mirage-like sensation—also associated with fog, as we shall see later—, what is emphasized in this first-contact scene is the immigrant's wish to make a home in this new land, his need to turn America into his genuine homeland. The mirage nature of the American Dream is somehow exorcized by the hopeful naivety of young Carlos.

The Filipino boy's arrival in Bulosan's *America Is in the Heart* bears some resemblance to the stories told in Kingston's second book, *China Men* (1980). Historically speaking, the events narrated by both Kingston and Bulosan were set in pre-WW2 America: Carlos is said to migrate in the 1930s and the father in *China Men* also travels to America around that time, since he has to go through Angel Island, which closed in 1940. In terms of genre, both books were published as overt or oblique biographies and both share a testimonial feel. However, *China Men* was published in 1980, thirty-four years after *America Is in the Heart*, and, for both historical and aesthetic reasons, Kingston's narration is much more ambivalent than Bulosan's. For one thing, in *China Men* Kingston includes different versions of the father figure, who seems to function as a Chinese everyman (a positive version of the stereotypical "John Chinaman"). The narrator explains how her "legal father," as she will call him, originally migrated from China to the United States via Cuba. In her calculated, postmodern ambiguity, the narrator suggests that, for this last leg of the journey, her father might have been smuggled in a box: "I tell everyone he made a legal trip from Cuba to New York. But there were fathers who had to hide inside crates to travel to Florida or New Orleans. ... I think this is the journey you don't tell me" (Kingston 1980, 48). She then tries to imagine the sea voyage her father keeps silent about: he pictures him as a stowaway, hiding in a claustrophobic crate. Surrounded by silence and complete darkness for hours on end, the immigrant "lose[s] his bearings" (49). During the journey, the father manages to survive not only the physical hardships of confinement but also the constant uncertainty and fear of being found.[11]

The ship finally docks in the New York harbor and the "illegal" father sees the Gold Mountain for the first time. When the smuggler motions the new immigrant to contemplate the harbor, the "illegal" father is "thrilled enough to see sky and skyscrapers" (52). He seems to be pleased as much by the natural as by the artificial: on the one hand, relieved and eager to enjoy the freedom of open air, the natural "sky," he is also impressed by the technological marvel of "skyscrapers." And yet, as the term "enough" suggests, there is much more to this new environment than sky(scrapers); in fact, what the smuggler really meant and what he finally pointed at was the statue dominating the harbor:

> A gray and green giantess stood on the gray water; her clothes, though seeming to swirl, were stiff in the wind and the moving sea. She was a statue, and she carried fire and a book. "Is she a goddess of theirs?" the father asked. "No," said the smuggler, "they don't have goddesses. She's a symbol of an idea." He was glad to hear that the Americans saw the idea of Liberty so real that they made a statue of it. (266)

The "goddess of freedom" is a polyvalent icon and as such it has been used in American literature. In "Middle-Class Asian American Women in a Global Frame" (2004), Sau-Ling Wong lists four symbolic uses that are visible in Divakaruni's and Minatoya's fiction: the statue as "the nation's gatekeeper"; or else, "as a female embodiment of the nation," America seen positively as a "generous nurturer"; also as the incarnation of America's "most cherished values, such as freedom, equality, and the opportunity to pursue happiness"; and, last but not least, "as a highly exportable symbol instantly recognizable by people all over the world" (184). While the first symbolic meaning, that of the country's gatekeeper, cannot be obviated, here it is shadowed by the third use, that of the embodiment of American values. It must be emphasized, however, that the abstract idea of freedom takes *material* form in a colossal, human-made artifact.

In this initial version of the immigrant father's arrival, therefore, America is first perceived and imagined through human-made artifacts (statue, skyscrapers) and, only secondarily, by natural elements (sky). This conclusion is reinforced by the fact that the father's contact with the materiality of the new continent—the earth beneath his feet—is mediated by an artificial substance, concrete: setting foot on American soil, he has to teach "his legs to step confidently, as if they belonged where they walked. He felt the *concrete* through his shoes" (266; emphasis added).

Next to the story of the stowaway or "illegal father," Kingston juxtaposes that of the "legal" father; however, even though she later assures the reader that this "illegal" version of her father's journey could not have happened, the entire description has a semblance of realism that belies the narrator's justificatory statement. Notwithstanding the narrator's overemphatic disclaimer—"Of course, my father could not have come that way. He came a legal way, something like this" (267)—the postmodern mode of the book neither confirms nor disputes the truthfulness of the story.

In the "legal" version of the father's voyage to America, he travels not to the East Coast but to the West Coast. As soon as his ship, full of Chinese emigrants like him, sails into the San Francisco Bay, they are all taken to the Immigration Station at Angel Island. It is no wonder that, during their long stay at the station, waiting to be examined and in constant fear of deportation, many immigrants fell into despair. Some detainees, as mentioned before, found in poetry an outlet for their frustration and anguish, and they contrived to carve anonymous poems on the walls of the detention building. Kingston tells us how the Chinese immigrants "wrote about the fog and being lonely and afraid" (1980, 270), carvings that would later be known as the Angel Island poems. Although these sojourners know they are "almost within swimming distance of San Francisco" (267), it seems to them as unbridgeable a gap as the whole Pacific Ocean that they have just crossed. At times, the fog prevents them from seeing the city in front of the island. Such is their disorientation that the immigrants start to feel that "San Francisco might have been a figment of Gold Mountain dreams" (270). After all, as the narrator explains, Gam Saan, the Gold Mountain, was nothing but a Chinese *invention*, the Cantonese being "people with fabulous imaginations" (302).

What I find particularly significant in the material description of this first-contact narrative is the recurrent presence of fog: "Night fell quickly; at about four o'clock the fog poured down the San Francisco hillsides, covered the bay and clouded the windows. Soon the city was gone, held fast by black sea and sky. The fog horns mourned" (270). From a realistic point of view, the occurrence of this natural element in this scene makes a lot of sense, since fog frequently haunts the Frisco Bay, where Angel Island is located. As Limerick notes, we cannot forget that this particular spot "served as a focal point in the Chinese discovery of the [American] landscape," often acting as "a direct mirror for emotion" (1992, 1029). This is evident in some of the wall poems collected in *Island*. In Poem 13, "Random Thoughts Deep at Night," signed by a certain "Yu of Taishan," the insular landscape, especially the dense clouds and fog, act as the gloomy objective correlative of the poet's pessimistic feelings: "In the quiet of night, I heard, faintly, the whistling of wind./ The forms and shadows saddened me; upon seeing the landscape, I composed a poem/ The floating clouds, the fog, darken the sky./ The moon shines faintly as the insects chirp./ Grief and bitterness entwined are heaven sent" (Lai et al. 1980, 52). Poem 17 similarly links the foggy island to the immigrant's heartache: "At times I gaze at the cloud- and fog-enshrouded mountain-front. / It only deepens my sadness" (Lai et al. 1980, 54). Likewise, Poem 26 conveys the detainee's impotence by emphasizing the inexorability of natural elements like the clouds: "My grief, like dense clouds, cannot be dispersed" (Lai et al. 1980, 54). Just as weather conditions cannot be changed, it is beyond the immigrant's power to put an end to his/her imprisonment.

Apart from wall poems like those quoted above, the editors of *Island* also reprinted and translated a longer poem, "Imprisonment in the Wooden Building," originally published in the *Chinese World* in March 1910. Like the wall poetry, this three-page narrative poem often resorts to natural images both to describe the reality of arrival and detention and to underscore the feeling of despair of the immigrant waiting season after season: "When I arrived in America, all I could do was gaze at the sea water in vain" (Lai et al. 1980, 139). Interestingly, the poem resorts to frequent animalization, as when the immigrants are imagined as fish caught in American (legislative) nets: "Together with several hundred countrymen, it is a slim hope to be the one *fish* to elude the net./ Half a thousand of the yellow race are here, feeling lost like *birds* in a fine, mesh net" (139). The second metaphor is even more powerful than the first, since the incongruity of birds being fished not only points at the absurd nature of the situation but also underlines the fact that freedom-loving creatures like birds or human beings should not be trapped and imprisoned. In the middle section of the poem, nonhuman nature is imagined as complicit with American unjust laws, in that neither heaven nor earth, clouds nor trees offer help to the immigrants. At best, the landscape, the animals, and other nonhuman elements can only act as passive mirrors of the poet's sadness:

> Being here, I wish to cry out to heaven, but heaven does not hear.
> Entering this room, I wish to call to the earth, but the earth does not answer.
> The trees are also gloomy outside the prison; a hundred birds cry mournfully.
> Clouds and mists enshroud the mountain-side; a thousand animals, startled, flee.
> This is what is known as living with trees and stones and roaming with the deer and wild boars
> Alas! Alas!
> The scenery evokes my emotions.
> Everywhere is desolation. (Lai et al. 1980, 140)

Both in the short poems written on the wall and in the long poem quoted above, fog and mists often seem to swathe the island where the detention center is located. The recurrent presence of *fog* obviously denotes the real weather conditions, but it also carries an additional metaphorical burden. The common figurative use of the adjective *foggy*, for instance, underscores the feeling of disorientation, often carrying the added connotation of danger. Together with confusion and bewilderment, the term also suggests a *vague, unclear* quality that perfectly fits the situation of these Chinese immigrants. As such, the terms *fog* and *foggy* best capture these people's own psychological state: confused by the incomprehensible language, fuddled by the new circumstances, subjected to demeaning inspection (poked and tested as if they were cattle), and mired in uncertainty about their future (Kingston 1980, 53). In addition, the disorienting, obfuscating connotations of *fog* turn it into a quasi-magical element. A feeling of unreality seems to envelop misty scenes.[12] It can also be argued that, in Kingston's postmodern *China Men*, the ambivalent, hazy atmosphere conjured up by the fog likewise reinforces the undecidable nature not just of narrative but of reality itself.

A literal—as opposed to figurative—understanding of the motif of fog in this first-contact story has larger implications than mere realistic accuracy. The blinding quality of real fog serves to remind us of our human, physical limitations. Here, it is useful to compare the two versions of the father's arrival in America. In the first description, that of the stowaway father, nature is eclipsed by human artifacts: skyscrapers, Statue of Liberty, concrete. Conversely, in the longer, "legal" version of the father's arrival, apparently more faithful to the historical records of Asian American immigration, it is the materiality of nature—sea, sky, but especially fog—that obliterates or swallows human construction, the city of San Francisco. Human *hübris* thus meets its punishment; we are chastised, but not entirely. Human language still manages to pierce through the blinding, disorienting fog, and the wind helps to carry the message: "Words came out of the fog, the wind whipping a voice around the Island. 'Let me land, Let me out. Let me land. I want to come home'" (270). Paradoxically, it is the human desire to literally touch the land that manages to break through the impenetrable wall of fog.

The relevance of the fog motif in Kingston first-contact narrative is confirmed not just by its presence in the Angel Island poems but also by the fact that it reappears later in another crucial Asian American novel of immigration, Gish Jen's *Typical American* (1991). Here, the character known as Yifeng—later renamed as Ralph Chang—arrives as a "legal" immigrant, a scholar whose utmost aim in life is to gain a doctorate and return to China, triumphant. True to his determination, Yifeng spends most of the sea journey studying and completely disregards what happens outside his books. He does not even pay attention to the first signs of land; the only icon he is looking forward to is the emblematic Golden Gate Bridge, which he sees as a symbol of liberty and progress: "That splendor! That radiance! True, it wasn't the Statue of Liberty, but still in his mind its span glowed bright, an image of freedom, and hope, and relief for the seasick. The day his boat happened into harbor, though, he couldn't make out the bridge until he was almost under it, *what with the fog*; and all there was to hear were foghorns" (Jen 1991, 7; emphasis added). Once more, the fog seems to erase or at least question the solidity of the actual bridge, much like the Angel Island inmates often had to imagine, rather than see, the city behind the fog.

It is particularly significant that, in his first contact with America, Yifeng only seems interested in human-made marvels, like the massive pylons of the bridge (5) or, once in New York, in engineering feats like skyscrapers:

in this "city of cities," Yifeng shows appreciative amazement at "its subway, its many mighty bridges, its highways," the pilings of the Empire State Building, the "roller coasters, Ferris wheels," and "eating factories" (8). He marvels at the human technology that makes those modern wonders possible: "Only he even saw these things ...; only he considered how they had been made, the gears turning, the levers tilting. Even haircuts done by machine here! The very air smelled of oil. Nothing was made of bamboo" (8). It is no surprise, then, that during the sea voyage, the first sighting of land goes totally unnoticed: "He studies in the sun, in the rain. ... He studied as the horizon developed, finally, a bit of *skin*—land! He studied as that skin thickened and deformed and resolved, shaping itself as inevitably as a fetus growing eyes, growing ears. Even when islands began to heave their brown, bristled backs up through the sea" (7; emphasis added). Scholarly Yifeng remains unaware of the human-like birth of the new land emerging before his eyes. It is the zero focalizer that, through a powerful personification, allows us to see what goes on around Yifeng, while the Chinese immigrant continues reading and studying. Much the same happens in his train ride across North America, from San Francisco to New York: "famous mountains lumbered by, famous rivers, plains, canyons, the whole holy American spectacle, without his looking up once"; Yifeng was only aware that he was passing those sights because of what the other passengers said (7). However, despite his pragmatic scholarly resolve, Yifeng cannot help himself and occasionally peeks through the window: "'I hear what other people talk,' he'd say—at least usually. Once, though, he blushed. 'I *almost* never take a look at.' He shrugged, sheepish. 'Interesting'" (Jen 8; emphasis in the original). Even though Yifeng seems to prefer American "oil" to Chinese "bamboo," he indirectly admits to being "awed"—or, in his understatement, "interested"—by the grandeur of American mountains, rivers, and canyons.

We will finally turn to a peculiar first-contact narrative, *Jasmine*, a novel published by the late Bharati Mukherjee in 1989. The homonymous protagonist of the novel is a young Indian girl who is "smuggled" into America, hidden, very much like Kingston's "illegal" father, under a tarp, "learn[ing] to roll with the waves and hold the vomit in" (Mukherjee 1989, 93). Both Kingston's father and Jasmine depart from Caribbean islands (Cuba and Grand Cayman, respectively) and attempt to reach the Atlantic shores of the United States. However, the materiality of the first contact with the American continent is approached in rather different ways in *China Men* and *Jasmine*. While the "illegal" father's arrival scene briefly revisits the iconic encounter with the Statue of Liberty as the icon of freedom, the description of the "illegal" entry in Mukherjee's novel both follows and reverses the classic convention of first-contact narratives.

To begin with, in *Jasmine*, the sighting of the American land is preceded by a contradictory sequence of sentences: "I smelled the unrinsed water of a distant shore. Then suddenly in the pinkening black of pre-dawn, America caromed off the horizon" (95–96). On the one hand, the use of the verb *carom* speaks of the hazardous game that Jasmine is playing by immigrating as a stowaway: the ball can bounce either way. On the other hand, although the initial reference to "unrinsed" smells does not seem auspicious, the last sentence seems to anticipate a glorious *dawn* after the under-the-tarp darkness that stowaways had been plunged into and had had to endure for such a long time. What follows, however, is far from promising: what finally emerges in the horizon is neither the iconic Statue of Liberty nor imposing nature, but a polluting human contraption, a nuclear plant. Rather than conjuring up pristine, Edenic beauty, this first-contact narrative proffers an uncensored picture of industrialized America, the machine polluting and blemishing the garden (Marx 1964).[13] Nothing seems to be less pastoral than this image of the New World swallowed by waste. And yet, Jasmine's arrival is told in such a way as to incorporate the traditional blueprint of first-contact narratives: "The first thing I saw were the two cones of a nuclear plant, and smoke spreading from them in complicated but seemingly purposeful patterns, edges lit by the rising sun, *like a gray, intricate map of an unexplored island continent,* against the pale *unscratched* blue of the sky" (95–96; emphasis added). For John Hoppe, this description is highly indebted to pre-colonial discovery narratives: "The map of the unknown continent etched into the sky can only remind us of the visions of the earliest European explorers, an echo reinforced by the 'seemingly purposeful' design she reads into it" (1999, 147).[14] In connecting the metaphor of the discovery map, in "binding the image of the waiting continent to the field of technology and its ambiguous promises," Jasmine's perception of the new land "is unmistakably inflected with futurity" (147). Hoppe's interpretation is tinged with a certain optimism that I do not share, especially because the description of the materiality of arrival is far from promising: "I waded through Eden's waste: plastic bottles, floating oranges, boards, sodden boxes, white and green plastic sacks tied shut but picked open by birds and pulled apart by crabs" (Mukherjee 1989, 96). If this first-contact narrative is infused with "futurity," it is a bleak futurity that is conjured up. We have moved from Bulosan's "exhilarating experience" of first seeing a paradise, pregnant with possibilities, to Jasmine's dismal experience of having to make her way "through Eden's waste." The only redeeming aspect of this anti-pastoral landscape is the fact that the sky remains "unscratched": "The first thing I saw were the two cones of a nuclear plant ... against the pale *unscratched* blue of the sky" (95–96). Yet, the overall picture is dismal and anti-Edenic, post-pastoral (Gifford 1999) rather than pastoral. The bleak description found in this arrival scene, with its emphasis on waste and pollution, can be read as a symptom of the advent of a "postnatural" era (Deitering 1992) and an example of toxic discourse (Buell 1998).[15] Not only that, but this toxic landscape is also a "landscape of technology" (Wilson 1992, 257) in which nuclear plants emerge as "unnatural" plants, a sad substitute for the real thing and a reminder that we are now faced with a new, transnatural nature.

In the foregoing analysis we have seen how the materiality of the encounter with the new land is depicted in two main ways in twentieth-century Asian American literature: characters are rather engulfed and overtaken by America's natural environment, or else impressed by the human-made and the technological. Whereas the prevalence of the nonhuman is more obvious in the earlier texts (Bulosan's, Angel Island poetry), the characters in later narratives tend to foreground the human-made (Jen's Yifeng, Kingston's stowaway father, Mukherjee's Jasmine). It may well be that American modernity, with all its seductions and perils, is wholeheartedly embraced by Jasmine, as Hoppe maintains (1999, 146–7). In Jen's *Typical American*, however, Yifeng's belief in the human-made wonders (engineering feats like the Golden Gate or the Empire State), the embodiment of the American Dream ethos that he enthusiastically adopts upon arrival, is finally defeated by stubborn reality. Nevertheless, in both novels the arrival scenes incorporate the natural environment in significant ways, either to grudgingly admit its beauty (*Typical American*) or to bemoan its anthropogenic degradation (*Jasmine*). Even on those occasions when "nature" is pictured as either unimportant or too fragile, it is still, inescapably, the first thing that *touches* the new immigrants, in all the senses of the word.

NOTES

1. A shorter version of this analysis was presented at the 29th MELUS conference, held in Athens, Georgia, USA, in April 2015.
2. See Perry Miller's homonymous book (1974), as well as the prefaces to Dana Phillips's *The Truth of Ecology* (2003) and Hans Bak and Walter W. Holbling's "*Nature's Nation*" *Revisited* (2003).

3. As Manu Karuka eloquently puts it, "A focus on continental imperialism can help push the lie of U.S. exceptionalism to rest. The frontier does not distinguish the United States from other nations. Rather it situates the United States in a context of imperialism" (2019, 172), an imperialist project that, contrary to what is generally held, was already at work prior to the 1898 Spanish-American War (2019, 185). See Madsen (1998) for a critique of American exceptionalism.

4. Not even the Puritan settlers remained immune to the allure of the American "exceptional" landscapes: much as they chose to focus on the perils of the wilderness, they were also amazed by the natural wonders. Of course, what was left out of this European picture of the American continent is the way the marginal others—women and ethnoracial minorities, most notably Native Americans—cooperated in or resisted the building of the myth of Nature's Nation (Kantor 2007).

5. In Chap. 7 of *The Culture of Nature* (1992), Alexander Wilson describes the political, economic, scientific/conservationist rationale behind the establishment of natural parks and reserves in the United States and Canada, as well as tracing "the changing culture of parkland" (231) in both countries, especially the shift in focus from "natural resources" to "natural heritage" (247), "away from large reserves and towards the microenvironments of a multiplicity of wild land types" (246). For a complementary history of the rise of "nature tourism" in North America, see Wilson (1992, 22–51). For a discussion of the problematic "erasure" of ethnic minorities, most notably Native Americans, from American National Parks, see Kantor (2007) and Oatman-Stanford (2018).

6. I will focus only on those characters from Asian American immigration narratives, including classics like Carlos Bulosan's *America Is in the Heart* or Maxine Hong Kingston's *China Men*, who arrive by ship or boat, rather than by plane, and whose first physical contact with the American land is not strongly mediated by human-made structures, such as airports. Asian American texts whose characters arrive by plane also feature interesting moments, from the famous airport scene in *The Woman Warrior*, where Brave Orchid meets Moon Orchid, to more recent novels like Lê Thi Diem Thuy's *The Gangster We Are All Looking For*.

7. While this Immigration Station operated, the procedure was the following: whenever ships arrived, "immigration officials climbed aboard and inspected the passengers' documents"; while those immigrants "with satisfactory papers could go ashore," the rest of the passengers, mostly Chinese, "were transferred to a small steamer and ferried to the island immigration station to await hearings on their applications for entry" (14; Lai et al. 1980).

8. Poems 13, 15, 17, 18, 23, 26, 38, 40, 58, 64. There are also a few poems in the Appendix that feature natural elements: 13, 14, 28, 38, 54.

9. For a full discussion of Asian American cultural nationalism, see Chap. 4. For a recent analysis of agricultural discourse in Bulosan's *America Is in the Heart*, see Wald (2016, 131–144).

10. Note the personification that further links the building to the white American humans who erected them.

11. When the smuggler finally allows him and the other Chinese stowaways to get out of their boxes, the "illegal" father does not lose this sense of fear and menace: he watches an ominous "trouser leg turn this way and that. He had never seen anything so white, the crease so harp. A shark's tooth. A silver blade. ... Then, blessedness, the trouser leg turned once more and walked away" (52).

12. This unreal or surreal atmosphere is typical of war narratives like Coppola's *Apocalypse Now* or Tim O'Brien's *The Things They Carried*. "The fog," writes O'Brien, "made things seem hollow and unattached" (18).

13. This emphasis on the artificial and technological, which effectively contributes to the detriment of America's Edenic discourse, is interpreted by Hoppe as a "subversive re-appropriation of the already exhausted American cultural narratives of newness, open possibilities, and unknown but promising futures by new, postcolonial immigrants" (1999, 147). For John Gamber, however, there may be some redeeming aspects in the symbolic wielding of "toxicity," a term he favors over those that "imply former states of purity," such as pollution or contamination (2012, 5). We will return to this issue in Chap. 6.

14. However, Hoppe emphasizes the fact that the interpreter of this map, as an illegal immigrant, is not authorized or sanctioned "by Western narratives of power and possession" (1999, 147).

15. The politics of waste that we found in the novel's pivotal scene, with its ominous nuclear plants and insidious waste, is reminiscent of that found in other novels published in the 1980s. Cynthia Deitering claims that, in that decade, there was a growing awareness among Americans that the "pristine" character of nature was no longer there, that we had entered a "post-natural" era defined by consumption and waste, as we shall see in more detail in Chap. 6.

REFERENCES

Alexander, Meena. 1996. *The Shock of Arrival: Reflections on Postcolonial Experience*. Boston: South End Press.

Bak, Hans, and Walter W. Holbling, eds. 2003. *"Nature's Nation" Revisited. American Concepts of Nature from Wonder to Ecological Crisis*. Amsterdam: VU University Press.

Buell, Lawrence. 1998. Toxic Discourse. *Critical Inquiry* 24 (3): 639–665.

Bulosan, Carlos. (1946) 2000. *America Is In the Heart.* Seattle: University of Washington Press.

Deitering, Cynthia. (1992) 1996. The Postnatural Novel: Toxic Consciousness in Fiction of the 1980s. In *The Ecocriticism Reader: Landmarks in Literary Ecology,* ed. Cheryll Glotfelty and Harold Fromm, 196–203. Athens: University of Georgia Press.

Gamber, John B. 2012. *Positive Pollutions and Cultural Toxins: Waste and Contamination in Contemporary U.S. Ethnic Literatures.* Lincoln: University of Nebraska Press.

Gifford, Terry. 1999. *Pastoral.* London: Routledge.

Hoppe, John K. 1999. The Technological Hybrid as Post-American: Cross-Cultural Genetics in *Jasmine. MELUS* 24 (4): 137–156.

Jen, Gish. 1991. *Typical American.* Boston: Houghton Mifflin.

Kantor, Isaac. 2007. Ethnic Cleansing and America's Creation of National Parks. *Public Land & Resources Law Review* 28: 41–64.

Karuka, Manu. 2019. *Empire's Tracks: Indigenous Nations, Chinese Workers, and the Transcontinental Railroad.* Oakland: University of California Press.

Kingston, Maxine Hong. (1976) 2005. *The Woman Warrior.* New York: Everyman's Library, Knopf.

———. (1980) 2005. *China Men.* New York: Everyman's Library, Knopf.

Lai, Him Mark, Genny Lim, and Judy Yung. (1980) 1993. *Island: Poetry and History of Chinese Immigrants on Angel Island, 1910–1940.* Seattle: University of Washington Press.

Law-Yone, Wendy. 1983. *The Coffin Tree.* Evanston: Triquarterly-Northwestern University Press.

Lê, Thi Diem Thuy. 2003. *The Gangster We Are All Looking For.* New York: Anchor Books.

Limerick, Patricia Nelson. 1992. Disorientation and Reorientation: The American Landscape Discovered from the West. *Journal of American History* 79: 1021–1049.

Madsen, Deborah. 1998. *American Exceptionalism.* Jackson: University Press of Mississippi.

Marx, Leo. 1964. *The Machine in the Garden: Technology and the Pastoral Ideal in America.* New York: Oxford University Press.

Miller, Perry. 1974. *Nature's Nation.* Cambridge: Harvard University Press.

Mukherjee, Bharati. 1989. *Jasmine.* New York: Fawcett Crest.

Oatman-Stanford, Hunter. 2018. From Yosemite to Bears Ears: Erasing Native Americans from US National Parks. *Collectors Weekly.* https://www.collectorsweekly.com/articles/erasing-native-americans-from-national-parks

Phillips, Dana. 2003. *The Truth of Ecology: Nature, Culture, and Literature in America.* New York: Oxford University Press.

Wald, Sarah D. 2016. *The Nature of California: Race, Citizenship, and Farming since the Dust Bowl.* Seattle: University of Washington Press.

Wilson, Alexander. 1992. *The Culture of Nature: North American Landscape from Disney to the Exxon Valdez*. Cambridge: Blackwell.

Wong, Sau-ling. 2004. Middle-Class Asian American Women in a Global Frame: Refiguring the Statue of Liberty in Divakaruni and Minatoya. *MELUS* 29 (3–4): 183–210. https://doi.org/10.2307/4141850.

CHAPTER 3

"Naturalizing" Asian Americans: Edith Eaton

If the people … snub me, Norah, I'll send you over to steal their flowers
(Otono Watanna/ Winnifred Eaton; emphasis added)

The first-contact narratives that allowed us to enter Nature's Nation took us on a journey into America in which we (re)discovered the "natural" and "human" symbols of the country, from the Statue of Liberty to natural parks like Yosemite. Epitomizing as they do "imperishable" America and "utterly wild" nature in popular culture, these natural parks play a central role in the all-important process of nation building.[1] In contrast to European notions of national territory, in colonial contexts such as the American one, exploration and mapping were both a precondition of and simultaneous with nation building, and natural landscapes became crucial to such nationalist projects. In colonized areas, the natural environment is perceived to have been "there" or "in place" before its human inhabitants arrived, or, to quote Rich Heyman (2010, 329), "territory precede[s]" inhabitation. The same is true, of course, of every region of the world, including Europe, but European explorers first and, later, invaders and settlers conveniently saw the newly explored and colonized lands as "terra nullius." In the United States, this state of affairs would last until the nineteenth century, following independence from the British metropolis, when the project of nation building pushed the issue of nativism to the fore.

© The Author(s) 2020
B. Simal-González, *Ecocriticism and Asian American Literature*,
Literatures, Cultures, and the Environment,
https://doi.org/10.1007/978-3-030-35618-7_3

This opening chapter teases out the ways in which the nativist discourse of the late nineteenth century influenced the life and work of the first Chinese American writer, Edith Eaton, who was caught between two conflicting paradigms of animalization: the racist love implicit in the Orientalist image of the "pet" and the blatant racism at work in nativist depictions of Chinese immigrants as invasive, filthy "pests." I will explore how Eaton's work criticized the animalization of Chinese Americans and deployed alternative strategies of "naturalization" in order to reclaim and rehabilitate their denied humanity.

1   NATIVE/NATIVISM

The *Merriam Webster Online Dictionary* defines "nativism" in two ways: as "a policy of favoring native inhabitants as opposed to immigrants" and as "the revival or perpetuation of an indigenous culture especially in opposition to acculturation." While the first definition corresponds to the movement of the same name that emerged in the United States in the nineteenth century, the second is much closer to the postcolonial concept of nativism (Ashcroft et al. 2000, 143–144). For much of American history, and certainly during the nineteenth century, it was not indigenous Americans—the true "natives" of the land—who took up the nativist flag, but Americans of white "Anglo-Saxon" origin.[2] Contrary to the logic of indigenousness, it was the descendants of these White Anglo-Saxon Protestant (WASP) settlers who were constructed as the real natives of the United States. The nativist movement had begun.

By the end of the century, nationalist and racist discourses had converged and crystallized into a form of "racial nativism" which combined traditional Anglo-Saxonism with a new "scientific" school of "race-thinking" spurred by Darwinism (Higham 1955, 133–136).[3] In the mid-1880s, in a climate of social unrest, the racial nativists started to direct their attention toward the larger influx of immigrants from eastern and southern Europe, who were soon constructed "as constitutionally incapable of assimilation" (140).[4] Nativists found a convenient "scientific" justification for their views in new developments within the field of natural sciences. In particular, they looked to the eugenics movement and a growing belief in the powers of heredity (151–152; see Stepan and Gilman 1991, 84). For eugenicists, "the immigration question was at heart a biological one," making the prospect of allowing "degenerate breeding stock" to come into the country one of unpardonable consequences (151).[5]

Racial nativists' fascination with eugenics was matched by an equally passionate interest in anthropology, a discipline that seemed to provide a scientific classification of race. It was the work of European anthropologists—Gobineau and, later, Ripley—filtered through the American patrician, Madison Grant, that finally allowed nativists to build a complex scientific rationale for their exclusionary racist discourse. In *The Passing of the Great Race*, published in 1916, Grant contrasted the naively optimistic belief in the force of environment with what he viewed as the much more real power of heredity. Backing up his claims with reference to the new science of genetics, Grant identified the two major threats to "native" Americans of colonial descent: race suicide and reversion (Higham 1955, 156–157). With Grant's work, nativism finally had "a systematic, comprehensive world view" that privileged race over any other factor (157). This was racism, pure and simple.

If, as John Higham claims, racial nativism did not raise its ugly head until the turn of the century, how to account then for the earlier anti-Chinese movement in the American West? In fact, Higham's thesis about the emergence of racial nativism is only tenable if we exclusively consider the nativist campaigns against eastern and southern European immigrants. If we widen the lens to include anti-immigration activism against non-Europeans, and specifically against Chinese immigrants, evidence abounds that this racist shift was already visible in much of the Western part of the country by the 1850s, as confirmed by the Rock Springs massacre in Wyoming in 1855.[6] Not only was sinophobia common among the American West Coast working class by the 1870s, but, more surprisingly, it was also latent in the apparently "progressivist" discourse of many late nineteenth-century intellectuals and activists, such as Charles Lummis, the editor of the Western journal, *Land of Sunshine*, where Eaton published most of her fiction. Such intellectuals might acknowledge the undeniable existence of a Chinese American community in the main cities of the West Coast, but they thought of it as a passing phenomenon that would leave no lasting impression "in the American grain," so to speak.

Therefore, the intellectual climate of the second half of the nineteenth century contributed to feeding popular prejudices against non-white communities. Following their entrenched conviction that cultural features were inherited along with physical ones, many white Americans believed that, even after several generations, the inscrutable Chinese (American) would remain an inassimilable Other.

2   NATIVE/NATURE/NATURALIZATION

According to the *Oxford English Dictionary*, both human and nonhuman beings can be considered "native to a particular place." When applied to people, the meaning of the noun "native" is often tied with that of birthplace; thus, a human "native" is defined as "(3. a.) A person born in a specified place, region, or country, whether subsequently resident there or not." Similarly, fauna and flora can also be considered "natives" of a particular land: "(7. a.) A plant or animal that is indigenous to a country, locality, or habitat, and not introduced by man."[7] The emphasis here is on the fact that, if something or someone is born somewhere, it/she/he is "natural" to that area, in contrast to that which arrives or is introduced later by means of human intervention.[8] In theory, then, a person may only be considered a "native" of a particular country when born in that territory. This is the case of the United States, where a child born in America of non-American parents automatically becomes an American. However, if natives are considered to be (naturally) from somewhere in accordance with *ius soli* or birthright, logic would dictate that non-natives cannot be transformed into natives; what is not "natural" to or "naturally" from a certain place cannot be "naturalized." Yet "naturalization," a term commonly used in French, Italian, and English, promises just that.[9] The paradox of a non-native "natural" would seem to have been resolved by the process of legal naturalization, whereby a "non-natural" becomes a "citizen," with all the rights and duties that go hand in hand with that status. This is precisely what happened, historically speaking, for millions of people who traveled to America and opted for "naturalization." That path was not open to all immigrants, however.

The Naturalization Act of 1790 explicitly prevented non-whites from becoming citizens, and, although the meanings of "white" have shifted over the centuries,[10] it was generally black slaves, Native Americans, and Asian immigrants who became the locus of difference. These people could never be considered natives of the American land (or their equals), not even after centuries of coexistence in the same country. In fact, in the nineteenth and early twentieth centuries, only the few Chinese Americans who had been born on US soil were citizens by right of birth. To prevent this small number of Chinese Americans from growing exponentially, as nativists feared, additional laws were passed that effectively curtailed further immigration from China and the establishment of families for Chinese immigrants already residing in the United States. Thus, the Naturalization Act that had rendered non-white Asians ineligible for citizenship in the eighteenth century was reinforced by the Chinese Exclusion Act of 1882 and the Johnson-Reed (Asian Exclusion) Act of 1924.[11] Together, these legislative measures were intended to secure the eventual demise of Asian American communities. Through a complex interweaving of social racism and legal and institutional measures, especially via "racial isolation and concentration," nineteenth-century America seemed to have succeeded in establishing "color castes" aimed at keeping America white and free of dark-skinned peoples (Lyman 1990, 151).

It may be argued that our current obsession with citizenship may not correspond to the Chinese immigrants' main concern in late nineteenth-century America. In Eaton's "Plea for the Chinaman," according to Mary Chapman (2012a), the author appears to echo a desire not so much for citizenship, as it was conceived of at the time, as for the end of economic marginalization. However, Eaton's view of Chinese immigrants as having become "natives of Canada" would seem to contradict this theory. Precisely in one of the Canadian articles that Chapman analyzes, "The Chinese Defended," Eaton exclaims: "Why, the Chinese are the pioneers of British Columbia; they are the true British Columbians, and it is they and not the whites who should be claiming privileges from the Government" (quoted in Chapman 2012a). Chapman's skeptical view of Chinese immigrants' interest in becoming American or Canadian citizens is further undermined by evidence from the events that followed the San Francisco Earthquake and Fire of 1906. Most municipal records were lost in the disaster that hit the city, thus giving a sizable number of Chinese residents the chance to claim that they were American-born and therefore citizens, even when this was not the case. Not only that but, in the decades that followed, many immigrants arriving from China purported to be descendants of those real or fictive Chinese American citizens. Those who had falsely claimed their filiation came to be known as "paper sons." At least in the early decades of the twentieth century, therefore, the desire for citizenship was very much real. And yet, regardless of the immigrants' expectations of or desires for citizenship at one time or another, the fact remains that American legislation made it virtually impossible for Chinese individuals who wished to stay in the country and become citizens. Asian Americans have historically been "naturalized as an other in America" and have always approached its territory "conscious of borders, of places where you cannot or should not venture" (Hayashi 2007, 152). By 1924, those forbidden areas could be found everywhere in America; the whole country had been placed out of bounds for those "outlandish" immigrants from Asia.

3   NATURALIZATION/ANIMALIZATION: PETS AND PESTS

One can safely claim that the history of Asian America originates in the nineteenth century, when many of the Chinese "sojourners" decided to settle permanently in the new land. This was also a key century in the development of a scientific formulation of racism that placed certain human groups (including ethnic minorities) at the bottom of a "scientific" pyramid whose apex was occupied by the "white race."[12] Just as zoological taxonomies established a hierarchical pyramid of animal species—the basis for what Peter Singer would later call "speciesism" (1975)—scientific racism created a classification based on intra-human differences, constructed as "races." Implicitly, and frequently explicitly, the lower tier of this aberrant human pyramid was made to approximate and even coincide with the higher tier of the animal taxonomy, thus culminating a process of "theriometamorphosis," that is, the theriomorphic transformation of human beings, for obviously racist purposes. The trope of animalization served to contain the perceived alterity of racialized others by rendering them non-human and depriving them of the human rights afforded only to "proper" human beings.

Racist animalization has been described as "the organized subjection of racialized groups through animal figures" and usually "involves contextual comparisons between animals (as laborers, food, 'pests,' or 'wildlife') and the bodies or behaviors of racialized subjects" (Ahuja 2009, 557). As many scholars have demonstrated, racist animalization proved especially useful in colonial and imperialist ventures. Colonialist, racist, and sexist discourses share a similar rationale, since all of them derive "their conceptual strength from casting, sexual, racial and ethnic difference as closer to the animal and the body construed as a sphere of inferiority, as a lesser form of humanity lacking the full measure of rationality or culture" (Plumwood 1993, 4). Prior to the era of New Imperialism, Europeans had associated racialized others with nature, either through the benign myth of the "noble savage" or its exact opposite, the vicious primitive. One way or another, as Frantz Fanon reminded us in *The Wretched of the Earth* (1961, 42), colonialist discourse tended to resort to "the bestiary." While benign naturalization was common in the depiction of Native Americans and crystallized into what Krech calls the trope of "the Ecological Indian," the myth of the bloodthirsty Indian also featured prominently in popular genres such as the Western. The equation of black-skinned Africans with either wild beasts or tamable brutes was also frequent in colonial discourse and became even more cruelly commonplace following the large-scale importation of African slaves to the American colonies.[13]

If racist theriomorphic rhetoric was instrumental in colonial discourse, it proved no less useful in postcolonial contexts like that of nineteenth-century America, where Chinese labor was both feared by the self-appointed guardians of real America and necessary for the project of building the new nation. Animalizing strategies had proved very convenient when dealing with other minorities and internal colonies, such as African American and Native American communities, since in both cases these peoples were depicted as "primitive," and thus literally and figuratively close to the natural world. However, it was more difficult to apply such naturalizing strategies to a civilization with ancient written records such as China's. In the case of Chinese diasporic communities, naturalization did not derive from the Other's purported intimacy or identification with the natural world, but from an excess of "degenerate" culture. American racists and nativists made a conscious effort to magnify the inherent immorality of Chinese culture, which corrupted human beings and debased them into animals. The Chinese other was thus subjected to a particularly malicious form of "naturalization" as vermin or pests, species that have historically been perceived as filthy and/or harmful to human beings. As an extension of this type of naturalization, Chinese immigrants were not only cast as an inferior species but also objectified as soul-less bodies through the reductive, degrading synecdochic trope of the Chinese "hands," to be used and discarded once its usefulness was past. Indeed, these "hands" were actively sought during the building of Transcontinental Railroad, in the mid-nineteenth century; however, it was just their Chinese hands/bodies that were temporarily borrowed, while their disembodied human dignity was altogether ignored.[14] When those commodified bodies were no longer useful for the "modernization" of America, they were to be "returned" to their proper, "natural" place. In the meantime, popular depictions of the undesirable Chinese imagined them as unclean pests or deceivingly friendly pets who would pounce on their masters—or, even worse, on "their masters' women"—if they were not controlled or expelled.

The popular press, influential though it was at the turn of the century, was not the only apparatus to resort to theriomorphic metaphors for its ominous depictions of the Yellow Peril or its less sophisticated racial slurs. The scientists and intellectuals who had provided the scientific rationale for racialism were also seduced by the powerful trope of animalization. It is no coincidence that the major proponents of racial nativism in America,

Davenport and Grant, were amateur or professional zoologists—Grant even presided over the New York zoo. At the same time that these men carried out meticulous investigations of the native fauna, they were drawing implicit or explicit parallels between the human and nonhuman species. It was for them a short step to imply or fully assert the beastly, animal nature of certain "races," in contrast with the "spiritual" or rational character of their own, "superior" one. It was only logical for them to glide from imagining animal menageries to envisaging human zoos.

3.1   Pets and Curiosities

As might be surmised from our foregoing discussion, there were two major forms of anti-Chinese sentiment in nineteenth-century America: on the one hand, the more obvious sinophobia, which perceived the presence of Chinese and Chinese Americans as an imminent danger, and, on the other, a misleadingly benign Orientalism, masquerading as sinophilia but ultimately depriving its objects of interest of their agency and humanity. As might be expected, the animalization of Chinese immigrants and their descendants permeated both forms of anti-Chinese sentiment. Chinese people were regarded either as objects of exotic curiosity ("pets") or as potential threats ("pests").[15] In both cases, Chinese immigrants were subject to a white gaze that had the power to examine them as curios, out of voyeuristic, ethnographic interest, or as suspects, out of nationalist zeal. In both cases, the "animalization of the oriental" managed to "patch … over the hole rent by the emergence of the phantasm of absolute alterity ascribed to the racial other" (O'Quinn 1999, 166).

Just as the American "freak shows" and museums of the mid-nineteenth century had presented Chinese subjects as specimens of otherness,[16] so now the growing enclaves of Chinese immigrants came to represent a sort of human zoo full of curious and dangerous "creatures." One way in which these "abject others"[17] were typically animalized was by hyperbolizing or defamiliarizing their sexuality. Asian women were subject to demeaning strategies that turned them into sexual objects, just as African women had been previously depicted as hypersexual rarities.[18] In his analysis of the Orientalist nature of popular culture in late nineteenth-century America, Todd Vogel notes how Chinatown guides attracted tourists with bizarrely exotic lures: "One guide promised sword fights over slave girls; another enticed young men to prostitutes with the promise that Chinese women have vaginas that run east to west instead of north to south"

(2004, 110). This alien sexuality not only increased the voyeuristic attraction among prospective American clients but also implicitly reaffirmed the nonhuman nature of such outlandish creatures.

Nowhere was this purportedly "benign" form of animalization more obvious than in the designation of Chinese railroad workers as "Crocker's pets."[19] While the origin of the term may well lie in jealousy on the part of other workers toward their Asian competitors, the epithet also carried affectionate, even sexual connotations. This combination of patronizing attitudes to Chinese people as harmless, lesser beings, and tender implications of endearment is best captured by the term "racist love," coined by Frank Chin and Jeff Chan in the 1970s.[20]

Orientalism's obsessive interest in the "mysterious East," with its emphasis on otherness, inscrutability, and quasi-inhumanity, carried at its heart the seed of its dialectical opposite: racist hatred. Although Orientalist love, or sinophilia, might have seemed a more benign form of understanding Asian people and cultures than the blatant sinophobia that construed the Chinese as dangerous animals or "pests," both phenomena shared a dehumanizing momentum that made it very easy to slide from fascination into fear, from love into loathing.

3.2   Pests: The Yellow Peril

By the turn of the century, the stereotype of the inscrutable Asian was a source of both fear and fascination, both serving as bait for curios gazes and insinuating a veiled threat. By the time the Chinese Exclusion Act was passed, in 1882, the perception of the Chinese in the American popular imaginary had largely shifted from pets to pests. This transformation is best captured by a political cartoon, "In the Clutches of the Chinese Tiger" (Fig. 3.1), originally published in *The Wasp* in 1885. The comic strip includes six scenes, starting with a white family who welcomes and feeds a small kitten whose ribbon identifies it as "Chinese cheap labor." Scene by scene, the comic strip shows the metamorphosis of the animal from a playful little kitten into a spectacular, dangerous tiger. The underlying message is one of warning: be cautious in your welcoming of and your dealings with these 'quaint pets' because they may not only steal your jobs, but also overthrow your entire country—just as the grown tiger in the last image, still donning his exotic Mandarin hat, topples the table and pounces on the people, killing them.

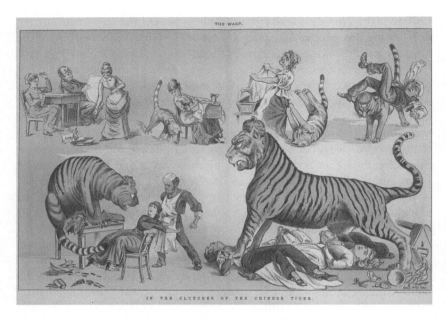

Fig. 3.1 "In the Clutches of the Chinese Tiger," *The Wasp*, November 7, 1885, p. 8

The movement from sinophilia to sinophobia, as depicted in the cartoon, coincided with the increasing visibility of Chinese immigrants in America at the end of the nineteenth century. As Yu-Fang Cho reminds us, while "the importance of the Chinese in U.S. imagination predates the first wave of mass migration of Chinese laborers during the Gold Rush in the mid-nineteenth century, the presence of the Chinese on the west coast in the last quarter of the nineteenth century remains a crucial moment in U.S. history of racialization" (2009, 39), not least because their demographic concentration in certain enclaves along the West Coast precipitated the emergence of markedly anti-Chinese sentiment and activism (Karuka 2019, 100). Though originally a Western states phenomenon, the sinophobic movement soon reached the ears of the federal government in Washington, where it acquired national relevance. This perceived threat at home combined with and fed off the international anti-Asian racist paranoia known as the "Yellow Peril."

Around the turn of the century, both Europe and the United States started to view the "yellow powers" of China and Japan as a serious menace, both economically and militarily. Crucial to the development of this "Yellow Peril" discourse was the circulation of fiction and essays by mainstream writers like Frank Norris ("The Third Circle"), Jack London (his 1904 essay "The Yellow Peril," as well as his 1919 short story "The Unparalleled Invasion"), and Olive Dilbert ("The Chinese Lily"), together with books on the geopolitics of China and Japan by self-appointed experts, such as Homer Lea, all of which seemed to sanction the fears fueled by the popular press.[21] As Fred Lee (2007) perceptively notes, the first publication to articulate the link between the international threat and the menace at home was Lea's *The Valor of Ignorance* (1909). Faithful to nativist discourse, Lea mourned what he considered to be the "decline of the Anglo Saxon in America" and its fateful consequences while speculating about an imminent Japanese invasion of California.[22] London's essay, "The Yellow Peril" (1904), also coinciding with the military rise of Japan, differed from Lea's by including both Japan and China in his analysis of "the menace to the Western world": "down at the bottom of their being, twisted into the fibres of them, is a heritage in common—a sameness in kind which time has not obliterated." The writer starts by apparently praising the hard-working, efficient nature of the Chinese he met when crossing from Korea into Manchuria: "Everybody worked. Everything worked. I saw a man mending the road. I was in China." Paragraph after paragraph, he repeats this refrain ("I was in China"), generally after some sort of eulogizing description of how things are done by the Chinese. After praising the Chinese people's endurance, hard work, and courage, London then goes on to extol the Japanese, whom he calls the "brown man." Here, London subverts the assumption that advanced, militarized Japan should constitute a more imminent "peril" on the basis of size: the country had but "forty-five millions," after all, so it could constitute no real danger for America. It was to China, London argued, that the Western world had to look instead, or, to be more precise, to a combination of Chinese skills and Japanese organizational power.[23] However, London hastened to add, we should not be afraid, because, while the Japanese might be able to imitate our technology and learn Western knowledge, the soul of the Western race, its spiritual endowment, could not be duplicated or acquired, but only inherited: "We are thumbed by the ages into what we are, and by no conscious inward effort can we in a day rethumb ourselves. Nor can the Japanese in a day, or a generation, rethumb himself in our image." Despite the final reassuring words, which echo the Eurocentric discourse of his times, London's message was clear: beware the yellow man.

While London was neither the first nor the only writer to regard the rising Japanese power with fearful admiration, he was unique in his flattering appraisal of Chinese potential. Historical analysts concur that the deployment of the "Yellow Peril" construct in America affected Chinese and Japanese immigrants differently. The sensational press of the time did much to feed the image of Chinese immigrants as filthy people who lived like animals, in contrast to the Japanese with their sophisticated manners. Accounts by several self-appointed experts in eastern life and philosophy appeared to confirm such a disparity. In *Seas and Lands* (1891), Edwin Arnold eulogized Japanese culture in sharp contrast to America's Chinatowns, which he described using the popular theriomorphic tropes of the time: the Chinese in San Francisco Chinatown, he wrote, "do not live in this extraordinary quarter, but rather wallow like pigs and burrow like rats," while at night they lie "packed … like sardines" (quoted in Ferens 2002, 43).[24]

Although the guises under which racist animalization functioned were many and multifarious, the pig/rat duo encapsulates most clearly the widely held view of the Chinese as dirty animals living in unhygienic conditions unfit for human habitation. For centuries, rats had constituted a threat to public health, not only because they were considered filthy but also because rats, the quintessential representation of harmful vermin, reproduced exponentially and could bring with them the most dreadful diseases. According to the racist discourse, the Chinese were rat eaters and, at the same time, they themselves were like rats. Additionally, the long Manchu queue (pigtail) worn by most nineteenth-century Chinese men provided racists with an easy pun on which to hang their equation of Chinese and nonhuman animals. Darwinian evolutionary theories were also invoked to justify their poisonous slurs, as shown in the following cartoons (Figs. 3.2 and 3.3) from the 1870s and 1880s. The pseudoscientific explanation presented in "Darwin's Theory Illustrated—The Creation of Chinaman and Pig," first published in *The Wasp* in 1877, traces the evolution of the Chinese from the monkey via the pig,[25] with the pigtail here representing the visible trace of animality linking all three figures.

In the second image (Fig. 3.3), a detail from a larger cartoon entitled "The Chinese Invasion," the animal tail is once again the visual constant in an equally disturbing metamorphosis. Keppler's lithograph, which first appeared in *Puck* in 1880, depicts the moment when the allegorical US

Fig. 3.2 "Darwin's Theory Illustrated—The Creation of Chinaman and Pig," *The Wasp*, January 6, 1877, p. 217

Fig. 3.3 "The Chinese Invasion," *Puck*, March 17, 1880, pp. 24–25

ship is starting to sink, and the only ones leaving and surviving the shipwreck are rats that gradually metamorphose into pigtailed "Chinamen."

Feline analogies could also be deployed when criticizing and animalizing the Chinese other, as seen above in the development of Chinese immigrants from seemingly harmless kittens into treacherous, dangerous tigers, and in authors such as Frank Norris: "In every Chinese there is something of the snake and a good deal of the cat" (quoted in Vogel 2004, 113). Sneaky, wily animals both, the comparison conjured images of a creeping, hidden danger. The Chinese might not show their fangs, but they certainly had them.[26]

4   EDITH EATON'S LIMITED OPTIONS

It is in this context of legal exclusion, Orientalist fascination, and xenophobic nativism that the first North American authors of Asian descent, Edith and Winnifred Eaton, began writing and publishing. Born in England in 1865, Edith Eaton moved with her family to the American continent at the age of five and spent the rest of her life in different parts of Canada (1870–1896, 1897–1898, 1913–1914), Jamaica (1896–1897), and the United States (1898–1913).[27] Even though her father was European and white, Edith signed her writings with the pseudonym, Sui Sin Far, in order to highlight the maternal Chinese heritage. This was by no means a necessary choice, as Dominika Ferens reminds us in her excellent *Edith and Winnifred Eaton: Chinatown Missions and Japanese Romances* (2002). In fact, Edith's writing career had started with pieces "on conventional subjects," and it was not until 1894, when she committed herself to "the home mission movement," that she became particularly interested in and vocal about the situation of Chinese immigrants in Canada and the United States (51).[28] With the advent of the "ethnic revival" and the recovery of multicultural literary traditions, Edith Eaton was soon claimed as a pioneer by Chinese Canadians and Chinese Americans alike (Ling 1990, 21; Doyle 1994, 50–51; Ling and White-Parks 1995, 2). There was no conflict of national allegiance for her, however: from the outset of her journalistic and literary career, she had been aware of and denounced the common plight of Chinese immigrants in both Canada and the United States. Influenced by the universalist ethos of Christian humanism, Eaton was not particularly interested in aligning herself with a single nationality: "After all I have no nationality and am not anxious to claim any" (1909, 230).[29]

Despite Eaton's open endorsement of the "Chinese cause," some critics remain skeptical of her representativeness as an Asian American.[30] Certainly, for writers living in turn-of-the-century America, like Eaton, Orientalism was a safe bet, and she was not immune to its temptations. Even her adoption of a Chinese *nom de plume*, Sui Sin Far, may be interpreted as an Orientalist move: by assuming a Chinese pen name, Eaton was actively marketing herself as an "authentic Oriental." However, it is no less true that the writer was aware of the risk of being turned into just another Oriental curio, and she consciously stopped short of fully assuming the Chinese persona that Orientalists demanded.[31] Eaton's choices must, therefore, be read in their proper context, taking into account both her personal background and the historical moment in which she lived. Both Edith and Winnifred Eaton "grew up surrounded by English Victorian culture in Canada, knew no people of Asian descent besides their mother, who had left China as a child, and their encounters with the Chinese and Japanese … were mediated by Orientalist ethnography" (Ferens 2002, 5). It is little wonder, then, to find these same patronizing attitudes toward Chinese culture mirrored in some of her own writings.

In order to reconcile and account for the contradictions in her work, we first have to realize that Eaton's embrace of artistic options that modern readers may well find unpalatable, politically questionable, or downright unacceptable was a conscious decision. This is especially true of her short stories, where Eaton's deployment of exoticism is more obvious. While her journalistic pieces show more political engagement with and sensitivity toward the sinophobia of the time, her fiction includes descriptions and value judgments that seem to follow the conventions and stereotypes of mainstream society. One of those conventions was sentimental literature, whose primary readership was female and often from the leisurely classes. Several critics have argued for a revision of the sentimental idiom, most notably Jane Tompkins in her groundbreaking *Sensational Designs* (1985). Pitting her analysis against the common appraisal of sentimental literature as trivializing and apolitical, Tompkins asks us to approach the domestic American novel of the nineteenth century "as a political enterprise, halfway between sermon and social theory, that both codifies and attempts to mold the values of its time" (126). She contends that the sentimental tradition "represents a monumental effort to reorganize culture from the woman's point of view" and, more importantly, "in certain cases, it offers a critique of American society far more devastating than any delivered by better-known critics such as Hawthorne and Melville" (124). Since Eaton's work was primarily meant for "a white middle-class audience that responded most readily to domestic themes in literature" (Ferens 2002, 12; see Tompkins 1985, 141), her use of sentimental conventions both pleased that audience and alerted them to the occluded humanity behind the Orientalist mask. Her stories revealed the real face of Chinese America: men and women who felt for their siblings imprisoned in detention centers, for the spouses they were afraid of losing, and for their children, who could be taken and kept away from them. Legal inequality was exposed and the fundamental equality of human beings was brought to the fore. By invoking the tremendous potentiality of the novel to foster change in readers and, indirectly, in society, Eaton's sentimentalism showed that the personal is political, that the affective can be effective.

A good example of the kind of appropriation of sentimental conventions highlighted by Tompkins and others may be found in Eaton's story, "Pat and Pan" (1912a), particularly in the depiction of the Chinese mother who adopts the white boy, Pat. This Chinese woman can be read as an embodiment of Victorian motherhood, "a conventional strategy of sentimental fiction" directly "appeal[ing] to readers' emotions" (Diana 2001). What is new in Eaton's fiction, however, is the conspicuous crossing of ethnoracial barriers in this adoption. This is a truly "revolutionary adaptation" of this sentimental idiom, a change that allows Eaton to demonstrate "the possibility of interracial love" (Diana 2001) and the value of transracial adoption. The fact that the author tampers with the conventions of sentimental literature in order to challenge or at least raise doubts about the rigidity of ethnoracial boundaries is evidence that she knew the potential of sentimentalism as well as its dangers.

Another feature of the sentimental genre was, as Min Hyoung Song (2003) puts it, its "popularization of a Christian morality, one that increasingly centered the example of Christ and his bodily suffering as the model of proper social behavior." It was hardly unusual, then, that Christian-raised Eaton should extol such virtues in her work, in stories like "The Chinese Lily," where the brother sacrifices romantic love so that his physically disabled sister can survive. This may be construed both as an analogical reference to Edith's own sacrifice for the sake of the weak (in her case, the racially othered) and as the exaltation of fraternal duty over romantic love, in keeping with traditional Chinese family values. At the same time, however, the sacrifice forced upon the brother's girlfriend might equally be seen as sanctioning the status quo, in an act of servile complicity typical of many sentimental novels.

Despite this valorization of the function of sentimental literature, Eaton's frequent recourse to tragic endings in which the Chinese become victims of other Chinese, usually as a result of a strict patriarchal system, only serves to reinforce Western views of the Chinese as a "degenerate race." Numerous scholars have attempted to explain why Eaton fell into this Orientalist trap by scrutinizing the popular demand for literary "exotica" and light entertainment among turn-of-the-century North American audiences (Ferens 2002, 63; Shih 2005, 67). While the articles that Eaton published in Montreal were very much informed by their local context and audience, those she contributed to American journals were influenced by the literary aspirations of their editors, so that "politics had to be muted, subject matter estheticized, and a certain degree of timelessness and universality assumed" (Ferens 2002, 67). On the other hand, the fact that Eaton often chose the tragic mode in her stories may also be seen as her own "way to counteract the pervasive tendency to ridicule and trivialize things Chinese" and, more importantly, to render them fully human in their feelings and affects (64). Unfortunately, Eaton's tragic endings not only attempted to capitalize on the power of sympathy but also seemed to mirror the fate awaiting the Chinese in America: disappointment, suicide, death, extinction.

It is only fair to point out, however, that Chinese and Chinese Americans fare much better in Eaton's writing than in the works of her contemporaries. Reading her work today, we must remember the "cultural noise" (Cutter 2002a, 35)—sensational journalism, self-appointed Orientalist experts, and photographic reportage, such as *Pictures of Old Chinatown* (1908) by Arnold Genthe, with text by Will Irwin—against which Eaton was attempting to make her voice heard. If the journalists, writers, and artists of her time resorted to Orientalism as a luring bait, Eaton tried to entice the readers with the promise of Oriental authenticity, but she used this bait and "tapped that prurient interest" (Vogel 2004, 105) for a different purpose. While Genthe and Irwin emphasized difference and exoticized the Other in their photographic representations of Chinatown, Eaton attempted to shrink the cultural distance between Chinese and American, and did so by imbuing many of her Chinese characters with conventional Victorian middle-class values—a price she was glad to pay in her crusade to demolish the constructed dichotomy between the two cultures.[32] Tempted as she might have been to confirm the Oriental stereotypes of the time, Eaton soon learned that this was a double-edged strategy: she would probably sell more, but she would alter nothing in the process. Thus, although Eaton needed to capitalize on her claims of authenticity to get some of her work published, her representation of the Chinese in North America at least tried to counteract and "counter-imagine" the simplistic and demeaning stereotypes of the time.[33]

In practical terms, Eaton's sense of mission was reflected in her efforts to show Chinese Canadians/Americans as complex human beings. Reversing the conventions of the day, Eaton allowed the Chinese other to occupy the symbolic center of her narratives, while white Americans were relegated to the margins and even to the position of the observed object.[34] Admittedly, in trying to bring the Chinese closer to a wider audience, Eaton does frequently echo the literary conventions of her time, particularly the regionalist and sentimental idioms of turn-of-the-century America, so eager for "local color" vignettes and stories full of predictable pathos. Even her Americanized Chinese characters, such as Mrs. Spring Fragrance, tend to exhibit a mixture of the familiar and the picturesque that was certain to appeal to readers of the time. Needless to say, most Chinese immigrants in North America did not lead such happy, easy lives.

For all that, there is much more to Eaton's work than clichéd Orientalism and predictable plots. Annette White-Parks, one of the first scholars to study the writer's life and work in depth, contends that, while indulging in a sort of "benign Orientalism" (1995a, 200), Eaton also resorted to "trickster stylistics" and played with expectations in many of her works, as in the inversion of stereotypes (White-Parks 1994, 1995b, c, 170). One of the triumphs of Eaton's work is the diversity and complexity of her depictions of Chinese and Chinese American people. Eaton endowed her characters with a human(izing) depth while revealing the diversity to be encountered among real Chinese immigrants. Consequently, in her fiction, she portrays an array of Chinese and Chinese American characters, ranging from the genteel generosity of the assimilated Mrs. Spring Fragrance to the shrewd and villainous Lum Choy in "A Chinese Ishmael" (1899), not forgetting the myriad of other personalities in between: demure, dainty ladies, or free-spirited girls like "Bohemian" Pan in "Its Wavering Image" (1912c), and traditional, authoritarian patriarchs, together with tender brothers and lovers as in "The Chinese Lily" (1912d) and "A Chinese Ishmael." Eaton's conscious commitment to multiplicity and diversity over dualistic depictions of good and bad (Ammons 1992, 117–18; Roh-Spaulding 1997, 159) stands as proof of her expertise in balancing white readers' (and editors') demands with her own sense of justice and loyalty to the "Chinese cause."

While some critics acknowledge a certain connivance in Eaton's work with the racist stereotyping of the time, most value her attempts to give voice and visibility to Chinese Americans. I believe that a critically sympathetic reading of her work is necessary in order to gauge the effectiveness of her use of Orientalism as a counter to anti-Chinese prejudices. It is my contention that Eaton's complex game of "evasion and disguise" (Roh-Spaulding 1997, 165) helped to dismantle the racist animalization of non-white minorities in America, by exposing the crude, dishonest fallacy of racialism and reversing existing "naturalized" stereotypes.

5   Eaton's "Fauna": Resisting Naturalization

One of the most interesting narratives created by Edith Eaton is her autobiographical "Leaves from the Mental Portfolio of an Eurasian," first published in 1909 in a New York-based journal called *The Independent*. In "Leaves," Eaton introduces us to the many instances of racism she encountered throughout her life, with the help of carefully crafted vignettes describing incidents that took place in Canada and the United States, and during her brief sojourn in the Caribbean. As we can see in this autobiographical narrative, many of the racial insults hurled at the young Edith and her siblings were of a dehumanizing tenor. Chinese immigrants in North America were perceived as less than human and, in consequence, more like animals. This discourse of dehumanization had a long history: from the beginning of American colonization by Europeans, non-white people were depicted as inferior beings, denied the soul and rational mind that would make them human. In the sixteenth century, Bartolomé de las Casas defended the cause of the Native American *indios* in the Spanish colonies by arguing that they had "buen intelecto" and "buena alma" (sic). His claims were refuted, however, by contemporary theologians and chroniclers (such as Gonzalo Fernández de Oviedo), who argued that Native American people's reasoning was "beast-like" ("entendimiento bestial"). In the centuries that followed, belief in the less-than-human nature of colonized peoples persisted among Westerners and was applied not only to Native Americans but also to people hailing from Asia and Africa who were not thought of as white. It is in this context that we have to understand Eaton's "Leaves."

In her autobiographical narrative, Eaton describes a pivotal dialogue that brought her face to face with racism. In one of her many jobs as a stenographer, Eaton attends a business dinner where she has to listen to demeaning comments from her companions. Upon passing a large group of Chinese railroad workers, Eaton's employer, who is not aware of her Chinese background, remarks: "I cannot reconcile myself to the thought that the Chinese are humans like ourselves. They may have immortal *souls*, but their faces seem to be so utterly devoid of expression that I cannot help but doubt" (1909, 224; emphasis added). For a committed Christian like Eaton, this was the ultimate insult: it deprived her, and anyone who shared her Chinese ancestry, of what was the essence of their humanity. In the Western philosophical tradition, the possession of a soul—and, later on, of its secular equivalent, Cartesian reason—established a "hypersepa-ration" between people and nature (Plumwood 1993, 49–51).[35] For centuries, Christian theologians had taught that it was the immortal soul that marked the irreconcilable difference between human and nonhuman beings. Raised as she was by English-educated, Christian parents, Eaton was deeply aware of the demeaning nature of animalization and its negation of any transcendental aspirations in the person or collective who were branded as "soul-less."

While traditional Western philosophy and Christian values continued to be important in Eaton's time, scientific discourse had gradually been eroding and displacing older philosophical, theological, and political discourses since the mid-nineteenth century. Fundamental ethical and political rights, such as the right to vote or to become "naturalized," were matters in which science was now an arbiter (Stepan and Gilman 1991, 81–82, 100). In addition, traditional Judeo-Christian worldviews were being challenged by new scientific discoveries, like Darwin's evolutionary theory in the field of biology. As we saw in our earlier discussion of racial nativism, slowly but relentlessly, Darwin's theories of evolution became socially sanctioned. As a result, simplified or bastardized versions of Darwinian theories in the field of biology were invoked and (mis)used to buttress racialist pseudo-science, and literature and popular culture explored and, in most cases, upheld this fallacy. Echoing the speciesist discourse of the time, racialists held that non-whites (which, for radical nativists, included immigrants from eastern and southern Europe) occupied a lower rung in the evolutionary ladder, closer to inferior species like animals (see Plumwood 1993, 4, 141). The use of animalization for racist purposes became pervasive in American popular culture and, for most of the nineteenth century Chinese immigrants were either pampered and paraded as pets or vilified as dangerous animals or undesirable pests. Being petted, or "basking in the sunshine of 'patronage,'" to quote Eaton (1895, 188), was tantamount to submitting to racist love. For cultural nationalists writing in the 1970s (Chin et al 1972, 1974), it would have been preferable for Asian Americans to suffer open aversion and racist hatred, rather than be symbolically castrated by such patronizing racism. At the time, however, there were those, like Eaton herself, who were not entirely "susceptible to such petting" (1895, 188). While critics differ in their analysis of the extent of Eaton's complicity with the Orientalist appetites of American culture, I believe that Eaton soon became aware of the distorting aspects of Orientalism and of how harmful this fascination with the Chinese other was to their personal dignity.

In "Half-Chinese Children" (1895) and "Leaves" (1909), Eaton resists and resents being turned not just into a spectacle but also into a dehumanized "creature." If the use of the term "creature" hints only indirectly at animality, other examples leave no room for doubt. Eaton narrates how, when she was a child, adults stared at her and her siblings as if they were "strange animals in a menagerie" (1909, 130). Similarly, several Chinese characters in Eaton's fiction begrudge the whites' "curious scrutiny" (1912c, 61), while many become aware that they are not valued as human beings but as specimens of a separate, nonhuman species (1895, 188–189). For Eaton, the powerless little girl in "Half-Chinese Children" is an obvious victim of this denuding, objectifying gaze: "they examine me from head to toe as if I was a *wild animal*" (1895, 189; emphasis added).[36] Although these examples still seem to correspond to the "pet" side of the pet/pest binary, the reference to the "wild" nature of the little girl in "Half-Chinese Children" hints at the future peril of the grown woman, just as the kitten-tiger cartoon discussed earlier (Fig. 3.1) signals the transition from domesticated pet to wild animal in the social imaginary of the day.

Even if the image of the Chinese as pets was still common in the late nineteenth century, by the time Eaton published most of her stories and essays, Chinese people were more often constructed as pests. Eaton was aware of this negative stereotype and, especially in her non-fiction, attempted to chronicle the racism she saw all around her. In her autobiographical narrative, Eaton recorded racist commonplaces, such as "[the Chinese] always give me such a creepy feeling," or her own landlady's ironic oath that "I wouldn't have one in my house" (1909, 224). In her short stories, Eaton resorted to more oblique strategies, whereby she denounced the unnatural disruption of families due to immigration laws, as well as the blind prejudices voiced by supposedly enlightened Americans. When the sinophobic animalization (and implicit dehumanization) of the Chinese as "disease-carrying aliens" or "malicious animals" (Cho 2009,

49) does seep into Eaton's writings, it never goes unchallenged. In other words, Eaton never internalizes the discourse of theriomorphic racism; she either attacks it or bypasses it by offering an alternative discourse.

At the time when Eaton started writing, as previously discussed, Chinese immigrants were commonly associated with low forms of animal life, such as rats and pigs. These two icons served the purpose of sinophobia: the former seen as invading, infesting vermin and the latter as despicable, dirty animals. It is no coincidence, then, that one refrain that Eaton persistently heard as a child and that she would later record in her autobiographical essay was "Chinky, Chinky, Chinaman, yellow-face, *pig-tail, rat-eater*" (219).[37] The image both highlighted tropes conjure up is one of abject animalization. Human beings do not have tails, unless, of course, you are Chinese. To be more specific, it is the pre-Manchu queue characteristic of Chinese male immigrants in the nineteenth century that refers synecdochically to the whole Chinese community. This apparently innocent fossilized metaphor also acts as a pejorative bond between Chinese immigrants and pigs, as illustrated by the pseudo-Darwinian cartoon (Fig. 3.2) analyzed earlier. An additional aspect of the racial rhyme that the children used to chant at Edith and her siblings is the way in which historically contingent characteristics (such as the queue or rat eating) were treated as essential and inherent features of "Chineseness." Not only that but, by eating rats, by introducing such an animal into one's body, one metaphorically becomes the animal and shares its abject status. Just as rats were seen as potential threats to the human body, so the Chinese were perceived as potential threats to the (American) social body.

Like rats, pigs also featured prominently in sinophobic discourse. Eaton's allusion to the pig slur is not only implicit in her explicit inclusion of the pigtail, it is also suggested in the metonymic references to animal enclosures, like pens, which once more bond Chinese people with "abject species" like pigs. Here, the pen becomes the figurative and literal site where pigs are kept and where Chinese immigrants were impounded upon entry into the country.[38] In one of her most popular stories, "Mrs. Spring Fragrance," Eaton subtly satirizes and criticizes the situation in which the Chinese Exclusion Act left immigrants hailing from China. In one key scene, a worried Mr. Spring Fragrance engages in a dialogue with his white neighbor in an attempt to make sense of the nature of love in the new country, fearing that his Americanized wife may not be happy in an arranged marriage such as theirs. During the course of this conversation, the Chinese man invites his younger interlocutor to a men-only "smoking party," since his wife, Jade, the Mrs. Spring Fragrance of the title, is away on a short trip:

"I wish to give a smoking party during her absence. I hope I may have the pleasure of your company."

"I shall be delighted," returned the young fellow. "But, Mr. Spring Fragrance, don't invite any other white fellows. If you do not I shall be able to get in a scoop. You know, I'm sort of an honorary reporter for the *Gleaner*."

"Very well," absently answered Mr. Spring Fragrance.

"Of course, your friend the Consul will be present. I shall call it 'A high-class Chinese *stag* party!'"

In spite of the melancholy mood, Mr. Spring Fragrance smiled.

"Everything is 'high-class' in America," he observed.

"Sure!" cheerfully answered the young man. "Haven't you ever heard that all Americans are princes and princesses, and just as soon as a foreigner puts his foot upon our shores, he also becomes of the nobility—I mean, the royal family."

"What about my brother in the *Detention Pen?*" dryly inquired Mr. Spring Fragrance.

"Now, you've got me," said the young man, rubbing his head. "Well, that is a shame—'a *beastly* shame,' as the Englishman says. But understand, old fellow, that *we real Americans* are up against that—even more than you. It is against our principles."

"I offer the real Americans my consolations that they should be compelled to do that which is against their principles." (1912e, 23, emphasis added)

In this excerpt, Eaton's criticism of Orientalist curiosity—the revelation that the young man only wants a "scoop"—is deftly juxtaposed with a condemnation of anti-Chinese legislation. When Mr. Spring Fragrance adds the light-hearted or ironic comment that "Everything is 'high-class' in America," the amateur journalist retorts that "all Americans are princes and princesses," in allusion to the fundamental equality of American society and its implicit contrast with the Old World, comprising both China and Europe, where nobility and class hierarchies still exist. However, Eaton cleverly inserts a cutting remark, so that the democratic spirit invoked by the young American in such a half-joking manner is belied by the immigration policies enacted by the American government. Mr. Spring Fragrance's brother, we learn, far from being treated as nobility, has been deprived of freedom and confined within the walls of an immigration station, referred to, significantly, as a "pen." Beyond the obvious denotations of imprisonment, the image of his brother's detention in this way contains more disturbing connotations of animalization.[39] The young man's concession that it is a "beastly shame" only underscores the rhetoric of animality suffusing the whole scene, starting with the initial allusion to a "stag" party. The final sarcastic sentence uttered by Mr. Spring Fragrance is a crushing indictment of American hypocrisy and shows the apparently old-fashioned Chinese man in a much more favorable light than the ultimately self-centered and condescending young American.

6   Eaton's "Flora": Benevolent Naturalization

In her bid to offer alternative images of the Chinese to emphasize their human nature and counter the sinophobia of her time, the paths available to Eaton were very few. Faced with racist animalization and objectification, Eaton could only allude to or offer some oblique criticism of the prejudice and discrimination suffered by Chinese Americans if she wanted her fiction to be accepted by the journals of the day. Nonetheless, it was still possible for her to deflate the power of racist discourse by co-opting its naturalizing strategies and employing them for her own purposes, just as she had done with the ethnographic gaze. Where naturalization constructed a demeaning image of Chinese people by associating or equating them with certain "repulsive" or "malicious" animals, an alternative naturalizing tactic could be wielded to counter those same negative representations. To the common racist strategy of abject animalization, Eaton responded with another potent element of nonhuman nature: flowers.

In the introduction to their 1995 edition of *Mrs. Spring Fragrance and Other Writings*, Amy Ling and Annette White-Parks admit that the stories' "somewhat *flowery* style" can be perceived by contemporary readers as *passé* (5; emphasis added). Eaton's stylistic choice is not only figuratively flowery, that is, elaborate or even pompous, but also literally full of flower tropes. Her unabashed preference for floral imagery was already apparent from her chosen pseudonym, Sui Sin Far, translated variously as "Chinese lily" (Lummis 1900), "narcissus" (Ling 1990, 41), or "water lily" (Doyle 1994, 50).[40] Her penchant for floral metaphors is also evident in many stories, where women are generally described as flowers or associated with them, most notably with the aforementioned Chinese lily. In the homonymous story, both female characters, Sin Far and Lily, bear the names of flowers and a flower metaphor is used to describe the Western convention of love at first sight: when Lin John encounters his sister's friend (Sin Far) for the first time, the sight of her calls to mind images of "apple and peach and plum trees showering their dainty blossoms in the country that Heaven loves" (1912d, 103). Similarly, in "A Chinese Ishmael" (1899), not only is the Chinese girl described as a lily, "a delicate little thing," but she also forgets temporarily her demure demeanor in order to throw Chinese lilies at the man destined to become her lover.

The visual paratexts accompanying Eaton's writings also incorporated stylized natural elements, usually plants. As David Shih perceptively notes, these illustrations, which were not chosen by Eaton herself but followed a general editorial policy, "operate as hermeneutic markers," so that, "even before reading the text, one comes to expect a narrative with 'Oriental' characteristics" (2005, 56). More pertinently for the purposes of my analysis here, these paratextual elements use natural tropes that most turn-of-the-century readers would have construed as positive, thereby helping to offset the negative, demeaning images often associated with Chinese people.

On the face of it, this apparently bland floral imagery might be dismissed as catering to an audience eager for Eaton's genteel, sophisticated version of the "Orient." However, it makes much more sense to interpret Eaton's choice as a familiarizing strategy designed to bridge the distance in popular culture between Chinese North Americans and "standard" Americans. To use Vogel's metaphor, Eaton "grabbed the edges of Genthe's [exoticizing] photos, expanded the margins, brightened the darkened spaces, and refocused the 'documentary' shots" (2004, 105–106) to show that the apparently unfamiliar was not so strange as initially imagined. This, however, had a problematic obverse, insofar as "writing about sameness does not sell—difference does" (Ferens 2002, 65). Inevitably, then, Eaton was obliged to scatter a few petals of exotica here and there.

The presence of gentle, flowery scenes in Eaton's fiction can also be read as a form of emulation. While perceptions of Chinese people in late nineteenth-century America were often negative, the Japanese were usually portrayed in a much more positive light. This disparity not only accounts for Winnifred Eaton's choice of a Japanese persona, but, in a rather unconscious way, provided Edith with an artistic solution to her dilemma. If, as Ferens demonstrates in her survey of the ethnographic literature of the time, the Chinese were generally seen as "wily" and the Japanese as "gentle," "quaint," and "dainty" (2002, 20, 30, 44; 20, 44; 27; 36), Edith Eaton only had to transplant what were perceived as

Japanese features onto her Chinese characters. One way of doing so was to adopt the sentimental mannerisms of Japanese romances, like her sister's, including their dainty little ladies, their flowery scenery, and the predictable dose of melodrama. Eaton's conscious use of the latter was determined not only by the popularity of the genre but also by her belief in its power to arouse the audience's sympathy to her own cause (see Ammons 1992; Tompkins 1985).

In particular, Eaton revisited nineteenth-century stylistic conventions regarding the use of natural metaphors and put them to different uses from the original ones. Thus, while "delicate" flowers are to be expected in the sentimental literature of the time, especially in exotic renditions of "Orientals," Eaton's use of flora imagery in her narratives both caters to the mainstream taste for the exotic and, at the same time, effectively counteracts the negative uses of abject naturalizing tropes, such as the instances of racist animalization discussed above. By tapping into the apparently facile style of sentimental romance, she managed to cater to mainstream tastes for the exotic while simultaneously constructing a viable alternative to the negative naturalization of Chinese subjects as pests to be eradicated. Her conscious use of popular conventions of late nineteenth-century sentimental literature made it possible for her audience to perceive Chinese immigrants in a more positive way. By proffering positive natural tropes in lieu of demeaning or dehumanizing ones, Eaton tried to counteract the sinophobic portrayal of the Chinese as irretrievably alien. Paradoxically, it was to the ultimate other, the nonhuman, that Eaton turned in her attempt to "fight back"; here, belying Audre Lorde's famous dictum, the (racist) master's tools are used to dismantle the (racist) master's house. Since in Chinatown even beauty seemed suspect, at least as filtered through Western eyes, Eaton went to pains to emphasize the inherently good nature of her Chinese and Chinese American characters, filling her most optimistic narratives with floral images of innocence and beauty.[41] To Norris's devious, wily snakes, Eaton responded with simple, faultless flowers, which were both a celebration of life and a symbol of purity in turn-of-the-century America.

In her reversal of the racist strategies of naturalization, Eaton complements the overwhelming presence of flowers in her stories with other positive ways in which Chinese characters interact with nature. Immigrants are depicted as enjoying wild nature, tending their gardens, or missing their animal friends. Even though most of Eaton's stories are set in urban Chinatowns, the natural environment is always present, either literally or metaphorically. "Tian Shan's Kindred Spirit" is an example of both: initially, the two main characters, Tian Shan and Fin Fan, cultivate an intense friendship in their frequent walks around a nearby Canadian mountain, and later on, when Tian Shan realizes he is in love with Fin Fan, it is animals that herald his epiphany: "the cat purred [Fin Fan's name]; a little mouse squeaked it; a nigh-bird sang it" (121). Similarly, in order to describe the surprise and disappointment felt by Fin Fan's father when his daughter rejects what he considers to be an advantageous arranged marriage, Eaton uses another animal metaphor: "Only a hen who has hatched a duckling and sees it take to the water for the first time could have worn such an expression" (122).

The natural rhetoric in Eaton's work tends to incline more toward the "domesticated" middle landscapes of the orchard and the garden, rather than the arresting views of the untamed wilderness. Nowhere is this preference more noticeable than in her 1897 article on Jamaican charitable institutions. When describing the great work carried out at the Industrial Farm School and Alpha Cottage, she likens the socially disadvantaged students attending both schools to tender plants to be nurtured. Like the latter, these seedling children have a promising future ahead of them: "I thoroughly enjoyed watching the streams of water flow over the land, and I think the little fellows enjoyed it too, for they laboured with hearty good will to divert the courses of the streams to the thirsty banana roots. Some of the trees were already bearing fruit. ... [A]ll the sixty-five boys on the farm are on their way to be men" (quoted in Ferens 2002, 76). As Ferens points out, "the reclaiming of these children is paralleled in the text by the children's successful reclaiming of 'what was but fourteen months ago a wilderness'" (2002, 76). Implicitly, Eaton seems to endorse the assimilation and domestication of these "wild" children.

Eaton's preference for domesticated middle landscapes, however, does not necessarily mark her as bourgeois. In her fiction, the leisure gardens and leisurely lives enjoyed by the likes of Mrs. Spring Fragrance are shown in parallel with the reality of other Chinese characters who eke out a living from their farms and market gardens. Eaton's primary interest in conventional domesticity, in families with parents and children, necessarily forced her to focus on the only Chinese American families that resembled the nuclear family paradigm, those of Chinese merchants, since, according to the restrictive legislation of the period, only they (and the occasional diplomat) were allowed to bring their spouses to the United States. Therefore, in her depiction of domestic "normalcy" among Chinese Americans, Eaton intelligently added all the middle-class trappings, including family animals (actual "pets") and gardens. The (Americanized) Spring Fragrances engage in all of the activities one would expect of genteel society: entertaining visitors, going for walks in the park, reading newspapers, and, most importantly, caring for their plants and animals. In "Mrs. Spring Fragrance," the title character writes to her husband and reminds him to take care of "the cat, the bird, and the flowers" (1912e, 22). The significance of this seemingly incidental instruction resurfaces later, when we learn that both Jade and her husband love their animal companions, especially the cat: "The cat jumped into his lap. He stroked it softly and tenderly. It had been much fondled by Mrs. Spring Fragrance, and Mr. Spring Fragrance was under the impression that it missed her" (1912e, 26). This apparently trivial detail confirms the Chinese character as a "sensitive" person, thus countering the insidious ethnography that maintained the "absence of nerves" (Ferens 2002, 30–31, 64–65) in people of Chinese origin. At the same time, the fact that Jade and, secondarily, her husband really care for their cat is rather ironic: this Chinese couple, tolerated and often "petted" by their white neighbors, have their own "pets," a cat and a bird. Thus, by including leisure-associated family animals in the Chinese household, Eaton makes it harder for readers to maintain their perception of the Chinese themselves as "pets." At the same time, the couple's "normal" domesticity, especially their affection for pets, humanizes them (Fudge 2008) by precluding any identification with debauchery, degradation, or the danger of "pests."

The issue of Eaton's limitations in her choice of characters is one that needs to be reassessed.[42] While the class and language barriers mentioned by some scholars were real, Eaton supplemented her limited first-hand knowledge with imagination and documentation, as any good writer must, and she included many working-class families in her stories. Also, while it is true that Eaton's preoccupation with domestic life led her to pay less attention to Chinese American bachelor society, it is no less true that some of her most intriguing stories feature single Chinese men who excel in their role as brothers or fathers, thus inverting the Victorian ideal of family while at the same time celebrating the domestic values of love, tenderness, and self-sacrifice. Examples of brotherly devotion and filial/parental love are found in "The Chinese Lily," "Lin John," and "O Yam—A Sketch," all of which, significantly, feature workers rather than wealthy merchants.[43] Even more importantly from an ecocritical point of view, stories like "O Yam" and "The Story of Tin-A" feature Chinese characters that work hard for a living and do so in an idealized middle landscape.

In "The Story of Tin-A," published in *The Land of Sunshine* in 1900, Eaton focuses on a family of Chinese farmers. When the tired, starving narrator arrives at the farm, the first thing she notices is the overwhelming fragrance coming from the garden: "The perfume of flowers stole to my nostrils from the plot of ground below the window, which was cram full of color" (1900a, 101). Sight follows the sense of smell and confirms an abundance of flowers, which in itself would not be a wonder in bountiful California, but for the suspicion of a loving hand behind this pleasant surprise: those flowers "gave me the impression of having been lovingly and patiently tended by one whose fondness for flowers was *inherent*" (101; emphasis added). The narrator soon discovers that the caring hand looking after the garden is that of an elderly woman named Tin-A. Self-figured as a non-Chinese outsider, the narrator confesses her curiosity about Tin-A and, in the ethnographic fashion of her time, ends up scrutinizing her object of interest, whom she describes as a "quaint little thing." The exoticizing adjective "quaint" reflects the narrator-focalizer's attitude, which corresponds to a benign version of Orientalism. On this occasion, however, the portrayal is not idealized but realistically frank: the narrator tells us that, for all her minute size, this old woman is "not as pretty as other Chinese women I have seen" (101). Tin-A's plain face is compensated, nevertheless, by the "goodness" that seems to envelop her. Even more interesting is the narrator's surprise at Tin-A's extensive "knowledge of the life of plants" (101). This is an interesting reaction: the white middle-class observer does not imagine Chinese women to be expert farmers or gardeners, especially at a time when only the wives of Chinese diplomats, scholars, or merchants were allowed into the country.

At the narrator's request, Tin-A reminisces about her happy childhood in Formosa—modern-day Taiwan—in a description steeped in pastoral rhetoric, and featuring the iconic Asian landscape: "My home was very beautiful. ... It was built on the side of a mountain which was ever green. Below our house and grounds were tea plantations and, further down, with trees and grasses lying between, were my father's rice terraces" (101–102). In that idealized landscape, young Tin-A and her best friend amuse themselves by "picking flowers" and "seaweed" (102). Tin-A, however, has to flee from this Edenic garden in order to escape an arranged marriage and avoid hurting her best friend. Her new companions, a group of Chinese actors, take her to America, where three of them end up settling as vegetable farmers. Tin-A makes the choice to leave behind her privileged life and becomes a gardener specializing in flowers, a detail which would also render her more agreeable to Eaton's genteel and bourgeois readers.

The brief "O Yam—A Sketch," first published in *The Land of Sunshine* in 1900, opens with a typical "picturesque spot" in the middle of Southern California where "tourists wandered amongst its flower-buried cottages" (1900b, 341). This floral exuberance prepares us to meet Wo Kee, a Chinese immigrant who, presumably prior to the passing of the Alien Land laws (see Chap. 4), "had come to the village, bought a piece of ground outside its limits, built a little shack, and started a market garden for the purpose of supplying the community with the succulent vegetables a Chinaman knows so well how to raise" (341). Wo Kee is portrayed as perfectly content in his Edenic garden, which thrives just as his little daughter O Yam does; the parallelism is no accident. Once more following the conventions of both sentimental and Orientalist literature, the child is first described as "a pretty little thing," whose "six-year-old cuteness quite captivated the hearts of the ladies" (341), a phrasing that echoes Eaton's own rhetorical strategy. We then learn that O Yam's personal charm is compounded by Chinese exotica: "a long braid interwoven with many colored silks hanging down her back and reaching almost to the heels of her tiny embroidered shoes" (341). Both her endearing daintiness and her quaint attire help to make her father's market garden a success among the lady customers.

This domestic bliss would have been all the more surprising for Eaton's nineteenth-century readers who, like one of the lady customers, were not used to thinking of "Chinamen" as good fathers:

> O Yam would not be *weaned* from her father's side for even one hour out of the twelve. There was only one person in the world for her, and that was her father. And Wo Kee's love for the child and his care of her were such that those whose knowledge of the Chinese was limited to books could not help but express surprise.
>
> "Ah, no," said Wo Kee one day, "not true that all Chinamen not care for girl child. Some think son better for honor family, and some too poor to keep girl, put her away, but parent-love always, boy or girl." (341; emphasis added)

The maternal-mammal metaphor used to refer to the father-daughter relationship only reinforces the tender nature of the father's affections. However, their domestic peace is often broken by external interference.

Wo Kee, "like all Chinamen living in America" (341–342), is also the victim of racist harassment:

> One day a number of [boys], passing his garden and seeing him there, began to pitch earth and pebbles on his back, at the same time making remarks on his dress and features. Wo Kee paid no attention whatever to his tormentors, but a *little creature* suddenly appeared on top of the garden fence, and with much childish dignity said:
> "Boys, foolish! has not my father a *spirit* that be much respect-worthy, and if that be so, what matter his face and his coat be not like yours? It be the *spirit*, not the nose, you ought love and respect."
> O Yam was then eleven years old, and though the boys laughed they could not help feeling small. (342, emphasis added)

Interestingly enough, here the Garden of Eden that is disrupted is not the white man's but the Chinese American's. At the same time, as White-Parks points out, "the narrator subtly reverses the usual fictional model that it is Chinese who introduce violence against White Americans," a reversal that she extrapolates to Eaton's readers: "We are with the Chinese-Americans inside their fortresses, experiencing their invasions from the surrounding wilderness by White Americans, the fictional 'Other'" (1995b). Anti-Chinese sentiment is once again exposed and contrasted with Chinese patience and forbearance. It is the little child's heroic love—much like that encountered in other sentimental narratives, such as H. B. Stowe's *Uncle Tom's Cabin*—that proves redemptive. O Yam's attitude, however, is far from submissive and complacent; instead, the girl launches a reasoned attack whereby the Chinese man's human dignity, symbolized by the term "spirit" (akin to "soul"), is vindicated. Constructs like "race" or "culture" mean nothing in the face of her father's inherent humanity. The child conveys her message through metonymic and synecdochic devices that disarm the older boys: it is not Wo Kee's "face" (i.e., his physical features) or his "coat" (his costume and customs) that matter; what marks him as human is his "spirit." Wo Kee remains silent during the scene; it is his daughter who bravely, even recklessly, articulates Chinese American collective anger.

That same strategy of juxtaposing Chinese forbearance and heroism reemerges at the end of the short sketch. The narrator tells us that Wo Kee has had an accident while traveling to San Francisco and lies dying in that distant city. The information is relayed to the child by "the telegraph operator's mother and sisters," who explain to O Yam that she cannot get to San Francisco any time soon and so will be unable to see her father before he dies (342). The quiet restraint with which O Yam receives the sad news causes the woman to marvel at the stoic nature of these "strange" Chinese people. The Orientalist cliché is belied immediately, however, by O Yam's brave determination to see her father before he dies:

> But even as they spoke a small hand plucked at their skirts.
> "I go see my father," O Yam said; and there was resolution in her voice.
> "Come home with me, poor little dear!" coaxed the old lady, taking O Yam's hand and seeking to lead her along. But the child would not be persuaded, and darted from her. (342)

O Yam is adamant that she will see her father one last time, and she does not hesitate to risk her life in order to do so:

> They were standing on a hill below which ran the railway track, and between the rails stood O Yam holding aloft a broom. Tied to the sweeping and upper end of the broom was a magenta silk garment—O Yam's best blouse. It fluttered in the breeze like a banner, and stretched itself out as if to greet the approaching train—not five minutes' distance off.
> "O Yam! O Yam!" the women screamed, clinging to one another.
> And to their straining ears was borne,
> "If I no see my father tonight, I no be live."
> They understood then; the child was risking her life to see her father die.
> "Good Lord!" cried one, "it is the fast express, and the chances are a hundred to one that it will go over her."
> The train thundered down. Its breath was on the child. The sisters covered their eyes, their mother fell on her knees murmuring a prayer. But the chance in a hundred was vouchsafed to O Yam. The train stopped—almost too late. And Wo Kee died that night with his little daughter's arms around him. (342–343)

Faced with such a melodramatic ending, modern readers may need to adjust their gaze and consider the context in which the story was written and published. At a time of intense sinophobia, Eaton chooses to highlight the heroic determination of a Chinese American child. In its celebration of filial love, the sketch champions the Chinese value of family duty and shows its proximity to the domestic morality of contemporary America. What is particularly significant about this story is the positive role played by natural images in the representation of the Chinese American family. This is nowhere more visible than in the naturalization of the main character, O Yam, who is identified with the plants and later linked with positive animal imagery: "she grew amongst Wo

Kee's asparagus, artichokes and vegetable marrows, as happy as a bird; trotting after her father as he worked around his garden" (341). Here, the Edenic rhetoric is interestingly revised by the positive naturalization of the Chinese American protagonist, who is not just *in* the garden but *part of* it.

Although floral tropes are more numerous and frequent, there are a few instances in which animal imagery is used to significant effect in Eaton's oeuvre. While her most common *nom de plume* was that of a typically Chinese flower, she also wrote under the pseudonym of "fire-fly," as in her writings in *Gall's Daily News Letter*, the Jamaican newspaper she worked for from December 1896 to the summer of 1897. More surprising for today's readers is the affection with which animals are portrayed in Eaton's fiction, particularly in "Mrs. Spring Fragrance," but also in "The Chinese Ishmael," "What About the Cat," "The Inferior Woman," and "Pat and Pan."[44] Nonetheless, their presence is not nearly as abundant or overwhelming as the "floral fragrance" that pervades her stories, starting with the title of her famous anthology. Through these narrative choices, Edith Eaton carried out the defiant advice voiced by her sister's alter ego, Onoto Watanna: if people "snub" you, go and "steal their flowers" (2003, 350).

7   THE QUANDARY OF RACIST ANIMALIZATION

The previous analysis has explored the way in which Eaton's work questions sinophobic stereotypes and attempts to deflate the anti-Chinese sentiment so rampant at the time of her writing. In order to make up for the contemporary dynamics of demeaning animalization, Eaton opts for alternative, positive modes of "naturalizing" Chinese Americans, new ways in which they can be associated with the natural environment. In a context where theriomorphism evoked the fear that, instead of Darwinian evolution, a Darwinian "devolution" (Vint 2010, 38) would take place, Eaton's choice of "harmless" flora over animal tropes allowed her to bypass contemporary anxieties about the erosion of the human/animal divide. Needless to say, it was primarily the obsession with "racial purity" that underlay the speciesist worries about the human/animal divide. Nineteenth-century thinkers, as Carrie Rohman notes, employed "the discourse of animality to articulate racist and sexist paradigms, projecting European species anxiety onto the framework of human difference" (2007, 28–29; see Plumwood 1993, 4, 107; Gamber 2012, 12–13). This was further complicated by the anti-miscegenation laws in America, regulations that equated interracial relationships with inter-species sex: something unspeakable and unimaginable, to be abhorred and avoided at all costs.

This slippage from speciesist to racist anxieties is thrown into stark relief in Eaton's "Its Wavering Image."[45] The story, which plays with and ultimately debunks the conventions of sentimental romance, revolves around a Eurasian girl who "'wavers' between a white and a Chinese sense of self" (Diana 2001), and whose name, Pan, underscores her ambivalence: "Like her namesake, the *hybrid man-beast* of the forest is caught between the conflicting forces of nature and culture" (Roh-Spaulding 1997, 170; emphasis added).[46] The choice of the faun to symbolize the plight of the child can give rise to certain misgivings. Who is the "beast" in the binary, we might ask. Is this a reversion to dehumanizing, theriomorphic racism? Or, on the contrary, is this subtle embracing of things "human" and "nonhuman" a commentary on the possibility of belonging to both realms?

These questions lead us to the logical quandary posed by racist animalization, the "utilization of animality in order to marginalize or distance the racial other" (Rohman 2007, 32). Animalization is rightly looked on as something to be shunned and deplored. The difficulty, particularly for ecocritics, is that the very act of unearthing and condemning instances of animalization that objectify and deprive "racial others" of their (human) dignity requires an implicit acceptance of human superiority over and distance from nonhumans. The danger, as scholars as diverse as Haraway, Plumwood, and Derrida have warned, is that "if we allow the human/animal distinction to remain intact ... then the machinery of speciesism and animalization will be available to use against various subjugated groups, animal or human, as history well shows" (Cole et al. 2011, 103).[47]

Those of us who are both critical race scholars and ecocritics find ourselves troubled and constrained by this predicament. Animalization has been insidiously deployed to perpetuate racism for centuries. If our analysis rejects animalizing tropes as pejorative, demeaning, and racist, it simultaneously appears to confirm that, if not the entire "animal kingdom," at least certain non-human species are devoid of all worth, an undesirable and unintended effect. Thus, any contemporary discussion of racist animalization must include a caveat, such as this, in which the evil purportedly inherent in certain species and the lack of value of nonhuman animals are radically questioned. The solution may lie in undertaking a reappraisal of the natural environment, interrogating and destabilizing the supposed worthlessness of certain species among our fellow biozens, at the same time that we continue to explore how "nature" has been used to subjugate and marginalize racialized others.

NOTES

1. For a description of how natural parks were constructed as devoid of human inhabitants, or at least of non-white presence, see Isaac Kantor's "Ethnic Cleansing and America's Creation of National Parks" and Hunter Oatman-Stanford's "From Yosemite to Bears Ears: Erasing Native Americans from US National Parks."

2. From our contemporary perspective it is difficult to deny that Native Americans were the "natural," native inhabitants of the land. That conviction was implicit in the historic events of the Boston Tea Party in the late eighteenth century, when one native people, the Mohawks, became the emblem and disguise used by the first decolonizers in their protests against British rule. With the establishment and consolidation of the new nation, popular imagination unraveled and reversed the logical equation that the natives of the land were the Native Americans.

3. According to Higham (1955, 5–10), three "main currents" fed nineteenth-century American Nativism: fear of "foreign radicals," which dated back to the late eighteenth century; anti-Catholic sentiment, coinciding with the arrival of the first Catholic (mostly Irish) immigrants; and "racial nativism," first fueled by an "Anglo-Saxonism" that did not differ much from its European counterpart.

4. Contradicting the more widespread thesis that it was the new immigration that fueled scientific racism, Higham maintains that "the new racial xenophobia did not originate as a way of discriminating between the old and new immigration. It arose from [social and economic] disturbances [...] that preceded the awareness of a general ethnic change in the incoming stream. At the outset, Anglo-Saxon nativism vaguely indicted the whole foreign influx. Only later did the attack narrow specifically to the new immigration" (1955, 137).

5. It was not only nativists but also writers and intellectuals who started to fear the influence of these new arrivals. In contrast to the earlier Darwinian optimism, there was mounting fear among intellectuals of the late nineteenth century that the new immigrants would eventually "outbreed" Americans of Anglo-Saxon descent (Higham 1955, 142).

6. Higham concedes that anti-Chinese sentiment predated the racialization of nativism, but he denies there is any connection between the two movements: "At no time in the nineteenth century did immigration restrictionists argue that Chinese exclusion set a precedent for their own proposals. The two issues were so different that foreign-born whites felt no embarrassment in leading the anti-Chinese crusade, while San Francisco's most bitterly anti-European nativists held entirely aloof from the war on the Oriental" (1955, 167).

7. Other elements of the environment could be considered "native," or "naturally" found in a certain place, although this meaning is now in disuse: "native: a metal or other mineral found naturally in a country or locality (*obs.*)" (*OED*).
8. The adjective "native" is also interesting and shares its etymology with the noun. According to the *OED*, the English adjective derives from

> Middle French, French *natif* belonging to the origin of an object (late 14th cent.), born in a particular place (early 15th cent.), (of metal) occurring naturally (1762; early 12th cent. in Old French (in a Franco-Occitan context) in form *natiz* in sense 'originating (from a place)') and its etymon classical Latin *nātīvus* having a birth or origin … innate, natural, naturally occurring, … in post-classical Latin also born in a particular place (9th cent.; late 12th cent. in a British source).

9. To "naturalize" somebody is to render him/her native. To quote the *OED* again:

> Etymology: < Middle French, French *naturaliser* (late 15th cent. In sense 'to make native'; mid 16th cent. in sense 'to adopt into a language'; late 16th cent. in sense 'to acclimatize') < *naturel* NATURAL *adj.* + *-iser* -IZE *suffix.* …
> To make native.
> *trans.* **a.** orig. *Sc.* To admit (a foreigner or immigrant) to the position and rights of citizenship; to invest with the privileges of a native-born subject.

> Neither Spanish nor Portuguese uses "naturalization" in this legal sense; however, Spanish and Portuguese (among other languages) use "natural" in the sense of "native." According to the *Diccionario de la Real Academia Española (DRAE)*, the adjective "natural" (from the Latin *naturalis*) means "2. Nativo de un pueblo o nación." Similarly, *Dicionário Priberam da Língua Portuguesa* explains that "natural" can mean "2. Oriundo, originário."

10. For an exploration of the construction of whiteness, see Richard Dyer's *White* (1997, re-edited in 2017) and Theodore W. Allen's two-volume *The Invention of the White Race* (1994, 1997). For a specific discussion of how East and South Asian immigrants were categorized, see Chan (1991) and Allen (1997).
11. "The Chinese Exclusion Act was only repealed in 1943 and naturalized citizenship for Asians was permitted in 1954, long after African Americans and American Indians were recognized as American citizens" (Ling 1990).
12. John Gamber concurs with other scholars that there is a "direct relationship … between racialized and anthropocentric hierarchies as well as the social and ecological dangers and cruelties they lead to" (2012, 13).

13. It is no coincidence that the most representative colonial(ist) novel of the turn of the century, Joseph Conrad's *Heart of Darkness* (1899), hinges on an "ethical regression" that relies "upon a deep ideological racism that conflates African and animal" (Rohman 2007, 29). See Outka (2008) for a study of the complicated relationship between African Americans and nature.

14. In the eyes of the American railroad barons, the status of Chinese immigrants working in the construction of the railway resembled that "of tools, of drills, and nitroglycerin," in Manu Karuka's apt words (2019, 89). For a discussion of "railway imperialism" or "railroad colonialism," see Davis and Wilburn (1991) and Karuka (2019), respectively.

15. The fact that American ethnographers could opt for "racial exclusion" or "cultural uplift," to use Lori Jirousek's terms (2002, 46), fits the pest/pet paradigm that I propose here. According to Jirousek, there were two types of "spectacle ethnographers": "those who feared racial degeneration through the Chinese and those who welcomed immigrant amalgamation as a means of Americanization" (2002, 29).

16. In *New York Before Chinatown* (1999), John Kuo Wei Tchen describes Peters's and Barnum's Chinese Museum, where a Chinese noble woman and her "suite" were exhibited in the mid-nineteenth century to great popular acclaim. Although museums and exhibits were not the only way in which the exotic was marketed and sold in the nineteenth century, and the Chinese became less of a rarity by the end of the century, the Western perception of "real Chinese" as curios continued to haunt readers like those that would come across Eaton's fiction and journalistic pieces. It is no coincidence, then, that her work both reflects and resents that exoticizing, objectifying gaze.

17. Vanessa Holford Diana (2001), following Judith Butler's theorization of Kristeva's abjection, has read the effective confinement of this "undesirable race" as the construction of abject "uninhabitable zones": "Occupying a defining role related to that of the "abject being," the Chinatowns of US urban centers served as the "uninhabitable zones" that Butler identifies as playing a necessary role in the act of racial differentiation and subsequent exclusion. She explains that this zone of uninhabitability will constitute the defining limit of the subject's domain; it will constitute the site of dreaded identification against which—and by virtue of which—the domain of the subject will circumscribe its own claim to autonomy and to life. In this sense, then, the subject is constituted through the force of exclusion and abjection. … The Chinese-American was constructed in mainstream imagination as embodying physical abjection and inhabiting the geographically unlivable."

18. The most obvious case was Sara Baartman, an African slave who came to be known as the "Hottentot Venus."

19. When talking about the demeaning strategy of racist animalization, I use the term "pet," with all its negative, patronizing connotations. Conversely, when referring to nonhuman pets, I favor phrases such as "animal friends," "family animals," "companion species" (Haraway 2003, 2006), and even "animal relatives" (Herrnstein-Smith 2004).

20. The perception of Chinese people as lesser beings, as pets, much like the obsession with "things Oriental" found among Orientalists, responds to the paradigm of "racist love" that Chin, Chan, Wong and Inada would decry in their work (Chin and Chan 1972; Chin et al 1974/1991). In both cases, Asians were rendered harmless, both "mysterious" and "quaint."

21. In May 1882, *Harper's Weekly* used their editorial piece to criticize the "Chinese Panic" promoted by certain political leaders and the sensational press. For a discussion of recent versions of sinophobic discourse, see Julie Sze's description of the new "China anxiety" (2015, 20–21, 26).

22. When writing this inflammatory volume, Lea drew on his own knowledge as a military advisor in China as well as on the new geopolitical reality that emerged from Japan's victories against Russia.

23. "The menace to the Western world lies, not in the little brown man, but in the four hundred millions of yellow men should the little brown man undertake their management. The Chinese is not dead to new ideas; he is an efficient worker; makes a good soldier, and is wealthy in the essential materials of a machine age. Under a capable management he will go far. The Japanese is prepared and fit to undertake this management. Not only has he proved himself an apt imitator of Western material progress, a sturdy worker, and a capable organizer, but he is far more fit to manage the Chinese than are we. The baffling enigma of the Chinese character is no baffling enigma to him."

24. For a survey of nineteenth-century Orientalist and missionary writings dealing with China and Japan, see Dominika Ferens's *Edith and Winnifred Eaton* (2002).

25. In another cartoon published in *The Wasp* on December 9, 1881, "The Chinese Question: The Remedy Too Late" (p. 400), it is the pigs' behinds that metamorphose into Chinese faces.

26. The closest to a fang that we get to see in the human representations in these cartoons is the exceedingly long nails of the Chinese figures. This would later become the signature trait of the Chinese villain *par excellence*, Fu Manchu. Other animals associated with Chinese immigrants were the octopus, the horse, construed not as a "noble" species but as a "beast of burden," and a plague of locusts (see Keller's "Uncle Sam's Farm in Danger," published in *The Wasp* on March 9, 1878).

27. For an overview of Eaton's life, see Ling (1990, 26–32), Ling and White-Parks (1995, 2–3), Ferens (2002, 185–187), and Chapman (2012b, 264–265); for a biography, see White-Parks (1995a).

28. It was not only the United States that was bent on annihilating the threat posed by the "Yellow Peril"; Canada implemented similar discriminatory laws and restrictions. Soon after the passage of this Exclusion Act in the United States, Chapman (2012a) reminds us, "the Canadian government followed suit, passing legislation in 1885 (after Chinese labour was no longer needed to build the transcontinental railroad) that required all Chinese immigrants to Canada to pay a fifty-dollar head tax, which increased in 1903 to five hundred dollars (or two years' salary)."

29. Nonetheless, Eaton was not immune to the identity quandary posed by the intractability of her "racial" categorization. While she apparently resolves that double allegiance by privileging her Chinese background, some critics continue to stress the tensions brought about by her own mixed ancestry; see Ling (1990); Ling and White-Parks (1995, 6); Doyle (1994, 52); Roh-Spaulding (1997); Leighton (2001); Diana (2001). Recent interpretations tend to view Eaton's in-betweenness as an asset or, in Chapman's words, as embracing "the thrill of not belonging" (2012a).

30. Vogel calls Eaton "a supplanter, who effectively 'reverse passes' from Anglo to Chinese," but who did so for a good reason (2004, 105). Ferens also concedes that Eaton "was not as close to Chinatown communities as once supposed and that even a sympathetic observer can feel a sense of detachment from or superiority toward the observed"; however, her empathetic involvement in the life of Chinese North Americans differs from the attitude found in her contemporaries (2002, 110). Maria N. Ng (1998) goes so far as to suggest that Eaton was a cultural outsider and therefore cannot really be viewed as a Chinese Canadian writer.

31. "They tell me that if I wish to succeed in literature in America I should dress in Chinese costume, carry a fan in my hand, wear a pair of scarlet beaded slippers, live in New York, and come of high birth. Instead of making myself familiar with the Chinese Americans around me, I should discourse on my spirit acquaintances with Chinese ancestors" (1909, 230).

32. Like her missionary mentors, Eaton tried to highlight the aspects of the inhabitants of Chinatown that brought them closer to American audiences and minimized those that exacerbated anti-Chinese sentiment. A good example may be found in her article, "Girl Slave in Montreal," which, despite the sensational title, does not pander to the popular fascination with "yellow slavery" (Ferens 2002, 59; 65–66). For a discussion of Western constructions of "yellow slavery," see Cho (2009).

33. New findings by scholars such as Martha Cutter (2002b, 2006), Dominika Ferens (2002), and, more recently, Mary Chapman (2012a, b, 2016) have expanded the size and range of Eaton's work in a substantial manner. Such archival work "has uncovered nearly two hundred additional texts of diverse genres, themes, styles, and politics published in more than forty different Canadian, United States, and Jamaican periodicals between 1888 and 1914" (see Chapman 2012a, b, 264). Many of these "rescued" texts correspond to the period 1904–1909, an obscure phase for which Eaton's first biographer, White-Parks, was unable to discover any publications by the author. During these years Eaton used different types of fiction in an attempt to adapt to the different periodicals in which she was writing, and the diversity of topics and characters is greater here than in her earlier period as a writer, "display[ing] a surprising range of viewpoints, including those of Native American women, US imperial adventurers, Philippine governors, self-supporting 'New Women' stenographers, and Japanese, Persian, and Arab children" (Chapman 2012b, 265).

34. Nowhere is Eaton's penchant for ironic inversion more noticeable than in the story "The Inferior Woman," where Mrs. Spring Fragrance announces her intention to write a book about Americans for her fellow Chinese to read: "Ah, these Americans! These *mysterious, inscrutable*, incomprehensible Americans! Had I the divine right of learning I would put them into an immortal book" (1912b, 33; emphasis added).

35. In her groundbreaking study, Plumwood argues that Cartesian dualism, which is usually credited with this mind-body, human-nature hyperseparation, is merely a reworking of its Platonic predecessor.

36. This phrase also appears in an earlier story, "Sweet Sin" (224). After the protagonist's death, the father, aware of his daughter's deep resentment of this treatment, resolves to throw her body into the sea lying between the United States and China, not only to signify her "in-betweenness" but also to prevent her from being made into a spectacle once more: "let her rest where no curious eyes may gaze" (225).

37. The image of the Chinese immigrants as rat eaters had been present in the satirical journals since the 1870s, the most famous example being George Frederick Keller's cartoon "Uncle Sam's Thanksgiving Dinner," published in *The Wasp* in 1877. See Thomas Nast Project for a comparative analysis of Keller's and Nast's versions of this peculiar Thanksgiving Dinner: https://thomasnastcartoons.com/selected-cartoons/uncle-sams-thanksgiving-dinner-two-coasts-two-perspectives/.

38. The presence of animal enclosures prefigures the animalization that Japanese Americans would suffer during WW2, as we shall see in Chap. 5. Not only were horse stables and pen-like barracks used during the Japanese American internment, but most of the animals invoked in sinophobic dis-

course reappeared in the anti-Japanese propaganda of the 1940s. Among other things, Japanese were said to "live like rats, breed like rats and act like rats" (quoted in Fujitani 2007, 29).

39. Statistically speaking, pigs are the most common form of "penned" livestock in both China and the United States (*The Economist Online*). The association of pigs and pens in this phrase is reinforced by the frequent use of the pejorative "pigtail Chinaman" at the time.

40. Eaton used different spellings when signing her stories and articles: Sui Seen Far, Sui Sin Fah, and, most frequently, Sui Sin Far, all of them "transliterations of the symbol for water lily" (Doyle 1994, 50). While many critics consider "Sui Sin Far" a mere *nom de plume*, for White-Parks it was not "a pseudonym but a term of address used by her family from early childhood," even though the name found on her birth certificate is Edith Maude Eaton (1995b). Ng maintains the opposite: Eaton's *nom de plume* was part of a "naming charade" and suggests "an invented identity" (1998, 180; 177). In a more comprehensive analysis, Joy Leighton also attributes her choice of name to an adult strategy rather than a family tradition (2001, 4). For an interpretation of Eaton's chosen name, see Roh-Spaulding (1997, 160) and Leighton (2001, 5).

41. On the metaphorical resonance of flowers and their ethnic and gender inflections, see Cutter (2002b), Miner (1991, 151–52). It is worth noting that among the genteel activities women engaged in at the time were the so-called charitable missions, one of the most active of which in San Francisco was the "Fruit and Flower" mission.

42. Numerous critics have highlighted Eaton's limited social scope. Although Eaton was a professional woman who had to work for a living, she was also part of genteel Victorian society, hence the "few references in her stories to working-class Chinese" and her penchant for "the merchant class or idealized laborers with middle-class aspirations" (Ferens 2002, 66; see 98–99). This can be explained because, for Eaton, what defined the Chinese was not so much the kind of work they did but their domesticity: "the ability to love parents, spouses, and children, as well as to maintain family while in exile" (2002, 66).

43. "Lin John" can also be read from an ecocritical perspective, especially since the brother's innocence is associated with the ineffable beauty of nature from the outset (1899, 117), while his sister's treason is couched in precisely the opposite terms. Thus, the American temptation of materialism is symbolized by the carcass of a beautiful animal, "a sealskin sacque," killed to satisfy human vanity (118–119).

44. In "The Chinese Ishmael," the two lovers who commit suicide metamorphose into two sea lions, animals which mate for life: "the spirits of Leih Tseih and Ku Yum have passed into a pair of beautiful sea-lions who wan-

der in the moonlight over the rocks, meditating on life and love and sorrow" (e-text, p. 49). Here animal loyalty is set above human frailty. While birds are common in Eaton's fiction, mostly as metaphors, there is one evident example of an extended metaphor associated with birds in "An Inferior Woman." In this story Mrs. Spring Fragrance acts as a bird-watcher, hiding behind a bush to observe and listen to the lovers in a park. This is further reinforced when, in her next conversation, she recites some lines about birds. Just a few days later, while she and her husband are sitting on their veranda, they see the sad lover, Will Carman, and Mr. Spring Fragrance tells his wife that Carman seems to have "failed to snare his bird." When the two lovers are finally and happily reunited, it is Mrs. Spring Fragrance who rejoices at seeing that, at long last, "Carman's bird is in his nest" (41).

45. For Roh-Spaulding, the "wavering image" of the story's title metaphorizes "Pan's own unstable image of herself" (1997, 172). However, as she also notes, the literal allusion to the "wavering image" occurs in connection with a rhyme/song in which a natural element, the moon, is paramount: Pan's white lover tries to assuage her doubts and "sings her a love song about the 'wavering image' of the moon casting its broken reflection upon the water–a symbol of heaven's perfect love cast imperfectly upon the Earth" (Roh-Spaulding 1997, 171). Diana reads this differently and emphasizes the romantic/erotic connotations of the moon: her perception, we learn later, has been blurred by the "wavering image" of the moon, symbolically linked to Pan's desire for Mark Carson and the temptation of his 'irresistible voice'" (2001, 64).

46. There is another "Pan" in one of Eaton's *Tales of Chinese Children*, "Pat and Pan," where we also meet a hybrid character. This time it is Pat who is "racially" white and culturally Chinese until his last betrayal. Yet, if we are to believe White-Parks (1994, 225), this story is as much about anti-miscegenation fears as it is about betrayal, and the avoidance of interracial marriage is intimately associated with the abhorrence of any "hybrid" offspring. For a discussion of such "pollution" anxieties, and of racist attitudes understood as "cultural toxins," see Gamber (2012).

47. The inability to escape this vicious circle affects both conservative and apparently progressive approaches (Haraway 2006, 100–101). What both liberal and conservative analyses of animalization share is the assumption of human superiority over and unbridgeable distance from the nonhuman. Both leave the human/animal or human/nature dichotomy fundamentally unchanged and "relatively intractable" (Rohman 2007, 32).

REFERENCES

Ahuja, Neel. 2009. Postcolonial Critique in a Multispecies World. *PMLA* 124 (2): 556–563.

Allen, Theodore W. 1994. *The Invention of the White Race.* Vol. 1. London: Verso.

———. 1997. *The Invention of the White Race.* Vol. 2. London: Verso.

Ammons, Elizabeth. 1992. *Conflicting Stories: American Women Writers at the Turn into the Twentieth Century.* New York: Oxford University Press.

Ashcroft, Bill, Griffiths, and Tiffin. 2000. *Postcolonial Studies: The Key Concepts.* New York: Routledge.

Chan, Sucheng. 1991. *Asian Americans: An Interpretive History.* Boston: Twayne.

Chapman, Mary. 2012a. The 'Thrill' of Not Belonging: Edith Eaton (Sui Sin Far) and Flexible Citizenship. *Canadian Literature* 212: 191–196.

———. 2012b. From the Archives: Finding Edith Eaton. *Legacy* 29 (2): 263–279.

———, ed. 2016. *Becoming Sui Sin Far: Early Fiction, Journalism, and Travel Writing by Edith Maude Eaton.* Montreal: McGill/Queen's University Press.

Chin, Frank, and Jeffery Paul Chan. 1972. Racist Love. In *Seeing Through Shuck,* ed. Richard Kostelanetz, 65–79. New York: Ballantine.

Chin, Frank, Jeffery Paul Chan, Lawson Fusao Inada, and Shawn Wong. 1974/1991. An Introduction to Chinese and Japanese American Literature. In *Aiiieeeee! An Anthology of Asian American Writers,* ed. Frank Chin, Paul Chan, Lawson Fusao Inada, and Shawn Wong, 3–38. New York: Mentor Books.

Cho, Yu-Fang. 2009. 'Yellow Slavery,' Narratives of Rescue, and Sui Sin Far/ Edith Maude Eaton's 'Lin John' (1899). *Journal of Asian American Studies* 12 (1): 35–63.

Cole, Lucinda, et al. 2011. Speciesism, Identity Politics, and Ecocriticism: A Conversation with Humanists and Posthumanists. *The Eighteenth Century* 52 (1): 87–106.

Cutter, Martha. 2002a. Empire and the Mind of the Child: Sui Sin Far's "Tales of Chinese Children." *MELUS* 27 (2): 31–48.

———. 2002b. Smuggling Across the Borders of Race, Gender, and Sexuality: 'Mrs. Spring Fragrance.' In *Mixed Race Literature,* ed. Jonathan Brennan, 137–164. Stanford, CA: Stanford University Press.

———. 2006. Sui Sin Far's Letters to Charles Lummis: Contextualizing Publication Practices for the Asian American Subject at the Turn of the Century. *American Literary Realism* 38 (3): 259–275.

Davis, Clarence B. 1991. In *Railway Imperialism,* ed. Kenneth E. Wilburn Jr. New York: Greenwood Press.

Diana, Vanessa Holford. 2001. Biracial/Bicultural Identity in the Writings of Sui Sin Far. *MELUS* 26 (2): 159–186.

Doyle, James. 1994. Sui Sin Far and Onoto Watanna: Two Early Chinese-Canadian Authors. *Canadian Literature* 140: 50–58.

Dyer, Richard. (1997) 2017. *White*. (2nd ed.) London: Routledge.

Eaton, Edith (Sui Sin Far). 1895. Half-Chinese Children. In Eaton 1995, 187–191.

———. 1899. "A Chinese Ishmael." *Overland Monthly* 34:43–49. Electronic Text Center, University of Virginia Library.

———. 1900a. The Story of Tin-A. *Land of Sunshine* 12 (January 1900): 101–103. https://archive.org/stream/landsunshine02unkngoog/landsun-shine02unkngoog_djvu.txt.

———. 1900b. *O Yam*—A Sketch. *Land of Sunshine* 13 (November 1900): 341–343. https://archive.org/details/landsunshine00unkngoog/page/n11.

———. 1909. Leaves from the Mental Portfolio of an Eurasian. *Mrs. Spring Fragrance and Other Writings*. In Eaton 1995, 218–230.

———.1912a. Pat and Pan. In Eaton 1995, 160–166.

———. 1912b. The Inferior Woman. In Eaton 1995, 28–41.

———. 1912c. Its Wavering Image. In Eaton 1995, 61–66.

———. 1912d. The Chinese Lily. In Eaton 1995, 101–104.

———. 1912e. Mrs. Spring Fragrance. In Eaton 1995, 17–28.

———. (1912) 1995. *Mrs. Spring Fragrance and Other Writings*. Urbana: University of Illinois Press.

Fanon, Frantz. (1961) 1963. *The Wretched of the Earth*. New York: Grove.

Ferens, Dominika. 2002. *Edith and Winnifred Eaton: Chinatown Missions and Japanese Romances*. Urbana: University of Illinois Press.

Fudge, Erika. (2008) 2014. *Pets: The Art of Living*. New York: Routledge.

Fujitani, Takashi. 2007. Right to Kill, Right to Make Live: Koreans as Japanese and Japanese as Americans During WWII. *Representations* 99 (Summer): 13–39.

Gamber, John Blair. 2012. *Positive Pollutions and Cultural Toxins: Waste and Contamination in Contemporary U.S. Ethnic Literatures*. Lincoln: University of Nebraska Press.

Haraway, Donna. 2003. *The Companion Species Manifesto: Dogs, People and Significant Otherness*. Chicago: Prickly Paradigm Press.

———. 2006. Encounters with Companion Species: Entangling Dogs, Baboons, Philosophers and Biologists. *Configurations* 14: 97–114.

Hayashi, Robert T. 2007. *Haunted By Waters: A Journey Through Race and Place in the American West*. Iowa City: University of Iowa Press.

Herrnstein-Smith, Barbara. 2004. Animal Relatives, Difficult Relations. *differences: A Journal of Feminist Cultural Studies* 15 (1): 1–23. Project MUSE.

Heyman, Rich. 2010. Locating the Mississippi: Landscape, Nature, and National Territoriality at the Mississippi Headwaters. *American Quarterly* 62 (2): 303–333.

Higham, John. 1955. *Strangers in the Land: Patterns of American Nativism, 1860–1925*. New Brunswick: Rutgers University Press.

Jirousek, Lori. 2002. Spectacle Ethnography and Immigrant Resistance: Sui Sin Far and Anzia Yezierska. *MELUS* 27 (1): 25–52. Literature Online.

Kantor, Isaac. 2007. Ethnic Cleansing and America's Creation of National Parks. *Public Land & Resources Law Review* 28: 41–64.

Karuka, Manu. 2019. *Empire's Tracks: Indigenous Nations, Chinese Workers, and the Transcontinental Railroad*. Oakland: University of California Press.

Lea, Homer. 1909. *The Valor of Ignorance*. New York: Harper and Brothers.

Leighton, Joy M. 2001. 'A Chinese Ishmael': Sui Sin Far, Writing, and Exile. *Melus* 26 (3): 3–29.

Ling, Amy. 1990. *Between Worlds: Women Writers of Chinese Ancestry*. New York: Pergamon.

Ling, Amy, and White-Parks, Annette. 1995. Introduction. In Eaton (1912), *Mrs. Spring Fragrance and Other Writings*, eds. Amy Ling and Annette White-Parks, 1–8. Urbana: University of Illinois Press.

London, Jack. 1904 (1910). The Yellow Peril. In *Revolution and Other Essays*. London: Macmillan. http://london.sonoma.edu/Writings/Revolution/yellow.html

———. 1919. The Unparalleled Invasion. In *The Strength of the Strong*, 60–80. London: Macmillan. http://london.sonoma.edu/Writings/StrengthStrong/invasion.html.

Lummis, Charles Fletcher, ed. 1900. *The Land of Sunshine*. Google. https://archive.org/details/landsunshine00unkngoog/page/n11

Lyman, Stanford. 1990. *Civilization: Contents, Discontents, and Malcontents and Other Essays*. Fayetteville: The University of Arkansas Press.

Miner, Madonne. 1991. 'Trust Me': Reading the Romance Plot in Margaret Atwood's *The Handmaid's Tale*. *Twentieth Century Literature* 37 (2): 148–168.

Ng, Maria. 1998. Chop Suey Writing: Sui Sin Far, Wayson Choy and Judy Fong Bates. *Essays on Canadian Writing* 65: 171–186.

O'Quinn, Daniel. 1999. Murder, Hospitality, Philosophy: De Quincey and the Complicitous Grounds of National Identity. *Studies in Romanticism* 38 (2): 135–170. https://doi.org/10.2307/25601385.

Outka, Paul. 2008. *Race and Nature from Transcendentalism to the Harlem Renaissance*. New York: Palgrave Macmillan.

Plumwood, Val. 1993. *Feminism and the Mastery of Nature*. London: Routledge.

Rohman, Carrie. 2007. On Marrying a Butcher: Animality and Modernist Anxiety in West's 'Indissoluble Matrimony'. *Mosaic* 40 (1): 27–43. JSTOR. https://www.jstor.org/stable/44030156.

Roh-Spaulding, Carol. 1997. 'Wavering' Images: Mixed-Race Identity in the Stories of Edith Eaton/Sui Sin Far. In *Ethnicity and the American Short Story*, ed. William E. Cain and Julia Brown, 155–176. New York: Garland.

Shih, David. 2005. The Seduction of Origins: Sui Sin Far and the Race for Tradition. In *Form and Transformation in Asian American Literature*, ed. Xiaojing Zhou and Samina Najmi, 48–76. Seattle: University of Washington Press.

Singer, Peter. 1975. *Animal Liberation: A New Ethics for Our Treatment of Animals*. New York: Random House.

Song, Min Hyoung. 2003. Sentimentalism and Sui Sin Far. *Legacy* 20 (1): 134–152. Project MUSE. https://doi.org/10.1353/leg.2003.0063.

Stepan, Nancy L., and Sander L. Gilman. 1991. Appropriating the Idioms of Science: The Rejection of Scientific Racism. In *The Bounds of Race: Perspectives on Hegemony and Resistance*, ed. Dominick LaCapra, 72–103. Ithaca: Cornell University Press.

Sze, Julie. 2015. *Fantasy Islands: Chinese Dreams and Ecological Fears in an Age of Climate Crisis*. Oakland: University of California Press.

Tchen, John Kuo Wei. 1999. *New York Before Chinatown: Orientalism and the Shaping of American Culture (1776–1882)*. Baltimore: The Johns Hopkins University Press.

The Chinese Panic. 1882. *Harper's Weekly*, May 20, 1882: 306–307.

The Economist Online. 2011. Counting chickens: Where the World's Livestock Lives. *The Economist*, July 20, 2011. https://www.economist.com/graphic-detail/2011/07/27/counting-chickens. Accessed 13 May 2017.

Thomas Nast Project. n.d. Uncle Sam's Thanksgiving Dinner – Two Coasts, Two Perspectives. https://thomasnastcartoons.com/selected-cartoons/uncle-sams-thanksgiving-dinner-two-coasts-two-perspectives/. Accessed 28 Sept 2016.

Tompkins, Jane. 1985. *Sensational Designs*. New York: Oxford University Press.

Vint, Sherryl. 2010. *Animal Alterity: Science Fiction and the Question of the Animal*. Liverpool: Liverpool University Press.

Vogel, Todd. 2004. *Rewriting White: Race, Class, and Cultural Capital in Nineteenth-Century America*. New Brunswick: Rutgers University Press.

Watanna, Onoto (Winnifred Eaton). 2003. A Neighbor's Garden, My Own, and A Dream One. *Good Housekeeping*, 347–53. Reprinted in *Onoto Watanna: A Half Caste and Other Writings*, ed. Moser Linda Trinh and Rooney Elizabeth, 109–121. Urbana: University of Illinois Press. http://www.jstor.org/stable/10.5406/j.ctt2ttcm8

White-Parks, Annette. 1994. We Wear the Mask: Sui Sin Far as One Example of Trickster Authorship. In *Tricksterism in Turn-of-the-Century American Literature: A Multicultural Perspective*, ed. Elizabeth Ammons and Annette White-Parks, 1–19. Hanover: University Press of New England.

———. 1995a. *Sui Sin Far/Edith Maude Eaton: A Literary Biography*. Urbana: University of Illinois Press.

———. 1995b. A Reversal of American Concepts of 'Otherness' in Fiction by Sui Sin Far. *MELUS* 20 (1): 17–34.

———. 1995c. Introduction to Part 2. In Eaton (1912), *Mrs. Spring Fragrance and Other Writings*, eds. Amy Ling and Annette White-Parks, 169–77. Urbana: University of Illinois Press.

CHAPTER 4

Thinking (Like a) Gold Mountain: Shawn Wong's *Homebase* and Maxine Hong Kingston's *China Men*

In 1913, one year before Edith Eaton's premature death at the age of 49, the state of California passed the Alien Land Law that changed the legal relationship between Asian immigrants and the land they had made their home. California was not the first state to curtail Asian immigrants' rights in this way: as early as 1859, Oregon modified its legislation to prevent Chinese immigrants from buying land; in its 1889 Constitution, the state of Washington prohibited Asians from owning land in its territory, and other Western states passed similar laws in the late nineteenth and early twentieth centuries. Although these Alien Land Laws affected nationals from all Asian countries, at the end of the nineteenth century most immigrants coming from Asia were of Chinese and Japanese origin, and it was they who were initially targeted by these restrictive regulations. The Chinese sojourners, as they were considered at the time, usually came from Southern China, principally from Guangdong province. When they left those southern harbors and headed for the American continent, most emigrants were looking for Gam Saan, the Gold Mountain.[1] Following the advice of a farmer-writer like David Mas Masumoto, who encourages people to dig up the roots of the physical place they inhabit and want to feel part of,[2] in this chapter I will offer an ecocritical reading of the imaginative relationship between the first Asian immigrants in America and the land they explored and, in most cases, settled in.

Asian immigrants, understood as a strategically constructed group (Sau-ling Wong 1993, 6), faced two separate restrictions in America

© The Author(s) 2020
B. Simal-González, *Ecocriticism and Asian American Literature*,
Literatures, Cultures, and the Environment,
https://doi.org/10.1007/978-3-030-35618-7_4

during the nineteenth and twentieth centuries: on the one hand, as discussed in the previous chapter (Chap. 3), they were soon confronted with laws that prevented them from becoming "naturalized" as Americans, and, on the other hand, they were confronted with measures that curtailed or effectively denied their right to own land. This limited access to citizenship and land property helped to shape the uniqueness of their situation as a minority and appears as a subtext in much of the fiction and non-fiction produced by Asian Americans in the 1970s and 1980s. While those legal restrictions are central to understanding Asian American writing as a whole, they are particularly relevant to the analysis of the relationship between literature and environment in Asian American narratives.

Not only does an ecocritical study of Asian American literature require a deep understanding of the history of the different Asian American communities, it also calls for a new methodological approach. "The experience of Gold Mountain," as Hayashi reminds us in his pioneering "Beyond Walden Pond," "is one separate from that of the Frontier, the Garden, or the Sublime" (2007, 64) and, as such, it obliges critics to rethink their old theoretical frameworks, or else forge new ones, in order to discuss this uncharted phenomenon. Part of this chapter will be devoted to reassessing the existing analytical tools and finding new ones where necessary, since new paths are sometimes the only way to discover and understand the new landscape before us. Two theoretical concepts will be explored in particular detail: land empathy and a process of inscription in the land that I have termed "inlanding." These critical tools have emerged from the realization that some Asian American texts exceed not only the most common ecocritical methods but also the very master narratives with which they have traditionally been associated in Asian American literary criticism.

If Maxine Hong Kingston's *The Woman Warrior* (1976) opened Asian American literature to both critical attention and academic standing, two other narratives were pivotal in consolidating that position: Shawn Wong's *Homebase* (1979) and Kingston's *China Men* (1980).[3] Both *China Men* and *Homebase* represent self-avowed attempts on the part of Asian Americans to "claim America" as rightfully theirs, thus endorsing one of the central tenets of Asian American cultural nationalism.[4] However, the enlisting of environmental strategies to further the Asian American cultural-nationalist project and its need to "claim America" was by no means a new phenomenon. The existence of politically active bioregionalist movements since at least the 1970s demonstrates that the conjoining of environmental awareness and nationalism/regionalism was new only in its application to Asian American writing.[5] Historically speaking, the linking of environmental concerns and regionalist or nationalist demands has proved extremely helpful (at least temporarily) when used by disenfranchised minorities to articulate their rights in relation to the land, either as indigenous inhabitants of the land or co-builders of the new nation.[6]

The coupling of environmentalism and the nationalist impulse in Wong's *Homebase* and Kingston's *China Men* sees Asian Americanness inscribed, not only onto an American official history that had done its best to erase all trace of it but also into/onto the American landscape itself. In *Homebase*, toponymic bonds help to reinforce the narrator's sense of belonging to America, yet the story inverts that logic of belonging by describing the visible imprint that Chinese Americans have left on the very territory they are claiming as their own: America belongs to the narrator and his ancestors because they have physically become the territory they are claiming. Although such literal bonding with the land reappears in Kingston's *China Men*, here the need to claim America is supplemented by a different, less conspicuous agenda which points at a less anthropocentric paradigm. To be more precise, what Kingston's book manages to do more openly than Wong's is to offer a glimpse of a new relationship with the land, a vision more akin to "land empathy" than to previous colonialist attitudes of domination. Before delving into these two books, however, we first need to understand the social and historical context in which they were produced, as well as literary conventions (complex pastoralism) and the prevailing discourse (cultural nationalism) that framed these texts.

1   Asian/American Land(scapes)

1.1   Asian Americans and the American Land: A Fragile Connection

If we accept the premise that "land sovereignty" constitutes the core of the environmental justice movement (Smith 2007, 190), the case of Asian Americans would be the most paradigmatic and salient example of the impossibility of "land sovereignty" in the American context. Throughout their history in the United States, spanning almost two centuries, Asian Americans have maintained a peculiar relationship with the American land. Unlike Native Americans, who had inhabited the North American landscapes for many centuries before the first Europeans settled in those territories, immigrants—especially recent ones—seemed to have a more fragile connection with the land. Even though their sizeable presence in continental America can be traced back to the mid-nineteenth century, Asian Americans only began to claim America as "rightfully theirs" late in the twentieth century. Until the emergence of Asian American cultural nationalism in the 1970s, most Americans of Asian descent were routinely perceived as not belonging on American soil. In contrast to the slaves uprooted from Africa, who had been forced to sever all ties with their continent of origin, many Chinese, Japanese, Korean, and Filipino Americans (to name just a few national communities) continued to have contact with their homeland and, in some cases, to travel back and forth between the United States and their countries of origin. Their status in the eyes of other (non-Asian) Americans thus wavered between that of a mere sojourner to be treated with curiosity or indifference and that of a danger-ous intruder, "the coming man."[7] In addition, for most of the nineteenth and twentieth centuries, Asian Americans were perceived not only as *homogeneous*—all of them "yellow" or "Mongoloid," in the racist termi-nology of the time—but also as *homogeneously alien*: "strangers from a different shore" (Takaki 1989).

In *African American Environmental Thought* (2007), Kimberly Smith contends that emotional attachment to a specific territory may be propiti-ated by legal attachment in the form of land ownership, with the problem-atic implication that love derives from possessing that which is loved. Conversely, lack of property ownership and citizenship rights effectively pre-vents disempowered communities from establishing "an emotional bond to the land," an argument that, according to Smith, is frequently found in African American environmental discourse (24). While this emotional con-nection with the land is taken for granted among "white" Americans, in the case of ethnic minorities "it had to be consciously created and fought for" (88, 89).[8] In attempting to extrapolate Smith's main thesis for her study of African American environmentalism to the Asian American case, I have worked with the hypothesis that Asian immigrants' tortured relationship with the land they helped to shape, literally and figuratively, but which could not legally belong to them, necessarily affected the lives of several genera-tions of Asian Americans and consequently haunted their artistic and cul-tural production. The general sense of loss and displacement pervading key Asian American narratives of the twentieth century, such as Kingston's *China Men* and Wong's *Homebase*, would seem to support that thesis.[9] Asians were, after all, the only immigrants in the twentieth century whose access to both property and citizenship was consistently denied not only de facto but also *de iure*.[10] Early in the century several land laws were passed that targeted East and South Asians, like the aforementioned California Alien Land Act of 1913, under which immigrants from Asia "could no longer buy agricultural land or lease it for more than three years"; in the decades that followed, the law's proscriptions were made more restrictive and more thoroughly enforced, and spread to other states (Chan 1991, 47). Arguably, then, first-generation Asian Americans, deprived as they were of the possibility of legally possessing the land where they had lived and toiled for years, could only create a tenuous, temporary bond with the land they were helping to shape, clear, and/or cultivate.

In addition, the exploitative conditions under which most Asian immigrants worked decisively altered the manner in which they viewed and experienced the new land. In her analysis of African American environmental discourse, Smith claims that exploitative labor generally "make[s] the workers less careful," and fosters in those agricultural laborers certain "ambivalence toward the land: a sense of alienation from the land and community coupled with a desire (and perhaps strategies) to come into a more meaningful and creative relationship with the natural world" (2007, 93).[11] Although no system of indenture can be equated with the "peculiar institution" of slavery, scholars have recently claimed that indentured, "racialized labor," even if formally paid, closely resembled the system of legalized slavery (Yang 2010, 75; Karuka 2019, 85–101).[12] Arguably, then, Asian immigrants' indentured labor and hard-working conditions fostered a similarly ambivalent attitude toward the land among workers, much like "the slave system that forced slaves into an intimacy with the natural environment but also tended to alienate them from it" (Smith 2007, 10). This ambivalence became textualized in Asian American literature, albeit in different ways, as we shall see in our reading of Wong's and Kingston's narratives.

1.2   Nature and Work: Asian Americans Changing the Face of the Earth

A narrow understanding of the environment as nonhuman "nature" and, more specifically, untouched wilderness may account for the neglect of Asian American texts by ecocriticism until fairly recently. As remarked by Robert Hayashi, this tendency to privilege pristine nature and "wilderness not only hides the nearly ubiquitous connection between human labor and the natural environment; it also silences the views of those who most impact the environment through their work" (2007, 65), immigrant farmers among them. Unsurprisingly, the overwhelming majority of authors studied by ecocritics in the twentieth century were Euro-Americans, mostly male and mostly in possession of enough leisure not to see their environment as their "work environment," unlike the majority of working-class and non-white individuals. Although proto-ecocritics such as Raymond Williams drew attention to the human sweat in the soil, so to speak, it was not until the twenty-first century that critics made a conscious effort to reconnect "nature" with human presence and labor.

When we think about the first literary depictions of human work in and with nature, our minds turn to different creation narratives from around the world: from the Aboriginal culture still present in modern-day Australia to the Judeo-Christian book of Genesis. However, if we are looking for a specific, consolidated literary genre that interweaves human labor and the natural environment, what probably comes to mind is the pastoral tradition, teeming with shepherds from the moment it originally emerged, in ancient Greece. The discourse of pastoralism was first dissected by (proto) ecocritics such as Joseph Meeker, Raymond Williams, and Leo Marx. Meeker's *The Comedy of Survival* (1972), to quote Kate Rigby's eloquent description, constitutes "[t]he first extended deployment of an ecocritical hermeneutics of suspicion to literature" and also the first open criticism of the pastoral mode, which Meeker considered "a form of escapist fantasy, valorizing a tamed and idealized nature over wild no less than urban environments" (2015, 127–28). In his groundbreaking *The Country and the City*, Williams explored the pastoral and counter-pastoral strands in English literature since its inception, observing that the idealized tone already found in the pastoral in Virgil's times (1973, 17) persisted in the English tradition, where it ultimately "served to cover and to evade the actual and bitter contradictions of the time" (45; see Phillips 2003, 16). While Meeker and Williams mostly focused on the dangers of the idealizing impulse behind the pastoral tradition, Marx's earlier *The Machine in the Garden*—an essential study of American pastoralism which we will return to in the last chapter—was more nuanced in its analysis and posited the basic distinction between positive and negative or, in his terminology, "complex and sentimental" uses of the pastoral (1964, 25). In more recent decades, Lawrence Buell has attempted to rescue the pastoral from one-sided attacks by emphasizing its slippery nature (1995, 44, 50), while Terry Gifford's *Pastoral* revisits Marx's "complex/sentimental" dichotomy, arguing that the term is not univocal but may be construed as either "oppositional or escapist" (1999, 44).[13]

Among the different studies of the pastoral mode in the Anglo-American context, I find Williams's pioneering book invaluable for this analysis. One of the central questions in *The Country and the City*, as might be expected of its Marxist author, is that of class. Williams's study focuses on the presence/absence of labor in artistic renditions of rural life in English literature. In particular, he bemoans the "magical extraction of the curse of labour," the denial of the "existence of labourers" implicit in country poems by Carew and Jonson (1973, 32).[14] The simple rural life commonly glorified in pastoral poetry concealed a far less amiable reality: "the social order within which this [subsistence] agriculture was practised was as hard and as brutal as anything later experienced" (37). For Williams, only a few English writers chose to focus on the real conditions of the rural environment, exemplifying a counter-pastoral "structure of feeling," as in Crabbe's *The Village* (1783). In the past, however, this counter-pastoralism had not received the same level of attention as the classical pastoral, according to Williams, who urged critics to "call the bluff" of earlier scholarship and unearth the frequently class-related material particularities that the idealized convention of the pastoral had elided for so long (18–19).

Such pastoral idealization, which necessarily excises human labor, reemerged later in rather naive forms of environmentalism. In "Are You an Environmentalist or Do You Work for a Living?," Richard White explores the difficult relationship between environmental criticism and modern work. According to White, the assumptions that underpin most ecocritical appraisals of human work in nature fall into two positions: either condemning any human interference with pristine nature or else "sentimentalizing" old methods of farming and agricultural work in a show of pastoral naivety (1995, 171). Setting out his stall as a committed environmentalist, White observes, "I am not interested in replacing a romanticism of inviolate nature with a romanticism of local work" (181). Nor is a separation of human labor and nature even possible: when we are at work, even merely typing at our computers, we are far from disembodied and we never cease to modify the natural environment with our activities.[15] Our apparent distance and disconnection from nature, White reminds us, is nothing but an illusion: "What is disguised is that I—unlike loggers, farmers, fishers, or herders—do not have to face what I alter. ... Most humans must work, and our work—all our work—inevitably embeds us in nature, including what we consider wild and pristine nature" (184–85).[16]

Unlike White and his fellow scholars and intellectuals, the first generations of Asian Americans were immigrant miners and farmers who worked
closely with the land and, more often than not, in a non-urban environment. In particular, for most of the nineteenth century, Chinese immigrants came from rural areas and peasant societies. Whatever jobs they
may have had in China, in America they became manual workers, at a time
when there was a high demand for unskilled labor. Both Kingston's and
Wong's narratives bear witness to the imprint left behind by the Chinese
immigrants who toiled on American soil, a territory that they marked in a
variety of ways. In what follows we will see how *Homebase* and *China Men*
negotiate the intractable relationship between work and the natural
environment.

2   SHAWN WONG'S *HOMEBASE:* CLAIMING AMERICA THROUGH INLANDING[17]

Native to China. Planted a century ago in
California's gold country where it now
Runs wild …. (Shawn Wong, *Homebase*)

As co-editor of the pioneering *Aiiieeeee!* anthology (1974), the poet
and novelist Shawn Wong was one of the first writers to raise "a shout of
resistance and triumph" on behalf of an Asian American literary tradition
(Partridge 2004, 91).[18] The stereotypical "aiiieeeee" of parodic representations of Asian characters, consciously re-appropriated by the editors of
the anthology, became an onomatopoeic metaphor for Asian American
anger at having been "long ignored and forcibly excluded from creative
participation in American culture" (*Publishers Weekly*). The polemical prefaces and introductory essays in the *Aiiieeeee!* anthologies were hailed as "A
Manifesto for an Asian American renaissance" (*Partisan Review*), or "cultural nationalism," as it came to be known in Asian American studies. In
these foundational essays, Frank Chin and the other *Aiiieeeee!* editors
called for a rejection of the "dual personality" myth, disavowed the common conflation of Asians with Asian Americans, denounced the emascu-
lating depiction of Asian American males in literature and the media,
criticized "white supremacy and Christianity" (1972/1974, xvi), attacked
Asian American authors who catered to racist white audiences, and
praised the very few who had resisted that "racist love."[19] Foremost among
the cultural or ethnic nationalists' concerns were the issues of visibility,

authenticity, rehabilitating Asian American masculinity,[20] and "claiming America." Despite its historical importance, however, Asian American scholars have criticized the project, both for its narrow definition of Asian Americanness and for making writers choose "between heroism and feminism," to use King-kok Cheung's apt phrase (1990; see also Cheung 2016, 29–66; Sohn et al. 2010). Nonetheless, even the detractors of the exclusive and androcentric nature of the ethnonationalist project acknowledge its importance, especially in the 1990s, when, as we shall see in the final chapter, the shadow of "denationalization" (Sau-ling Wong 1995) was starting to haunt the field of Asian American studies.

Wong's *Homebase*, published in 1979, was clearly informed by the *Aiiieeeee!* spirit. Through the novel's Chinese American protagonist, Rainsford Chan, and his personal quest, Wong explores many of the concerns of the *Aiiieeeee!* group in the 1970s. The novel lambasts social and institutionalized racism, rails against the pernicious effects of Orientalist mystique, and criticizes the easy conflation of Asians and Asian Americans, as illustrated by young Rainsford's confession: "I have no place in America, after four generations there is nothing except what America tells me about the pride of being foreign, a visitor from a China I've never seen, never been to, never dream about, and never care about. Or, at best, here in my country I am still living at the fringe, the edge of China" (Wong 1979, 66). Prompted by this realization, the protagonist consciously sets out to "claim America" for himself and his community, a quest especially popular in the era of identity politics, when espousing a certain "essential" identity became a precondition for the recognition and survival of an ethnic community.

The narrator-protagonist, fittingly named after the Californian town where his great-grandfather lived but which no longer exists (1–2), is a fourth-generation Chinese American youth, struggling to find his roots in America. In order to do so, he looks back along his family line, from his great-grandfather to his deceased parents, all of whom can offer "ancestral" help. While attempting to unearth the traces that his grandfather and great-grandfather have left in America, he conjures up memories of his long-dead father in order to try and understand his own self. Rainsford's quest for "roots" leads him to engage in real and imaginary journeys.[21] Through his family's diaries, stories, and letters, and through his own dreams and travels, Rainsford "move[s] across America picking up ghosts" (28). These ghosts tell him the history of his community and the book engages in what Elaine Kim has described as a "triumphant reaffirmation of the Chinese American heritage" (1982, 194). Thus, Rainsford learns of the hard work in the Sierras, where the first Chinese Americans built the Transcontinental Railroad, although the feat remained unwritten in the history books. He also discovers that his grandfather was a "Chinese vaquero" (47) and, with the help of a Native American stranger, finds out about Angel Island, where Chinese immigrants were held and interrogated (86–94). Finally, he learns of the existence of paper sons. Reminiscing about his grandfather, at one point Rainsford ventriloquizes and takes on the old man's voice: "I am my grandfather come back to America after having been raised in China. My father is dead so I've had to assume someone else's name and family in order to legally enter the country. ... All my sons after me will have my assumed name" (88). Since "identity is a word full of the home" (27), as Rainsford himself explains, the novel chronicles this search for a "home base," which is finally reached in the last chapter. The novel closes with the realization that Rainsford belongs in and to America, but, more importantly, that America belongs to him. In other words, the narrator ultimately claims America for himself and his ethnic community. Although the ghosts of history have helped Rainsford construct his identity and buttressed his sense of belonging to/in America, it is the land itself, the natural environment, that proves his most valuable ally in his project of claiming America.

2.1   *Nature as Metaphorical Trope*

An environmentally inflected reading of Wong's *Homebase* soon reveals that, beyond the toponymic associations scattered throughout the narrative,[22] evocations of the natural environment abound: tropes deployed to construct and understand Chinese Americans, literal collaboration in landscape creation, and body-landscape fusion. The first of these ecoliterary strategies appears on the opening page of the novel, where a species of tree, *Ailanthus*, is used metaphorically to stand for Chinese Americans:

> AILANTHUS altissima *(A. glandulosa)*.
> Tree-of-Heaven. Deciduous tree. All Zones.
> Native to China. Planted a century ago in
> California's gold country where it now
> Runs wild ... Inconspicuous greenish flowers
> Are usually followed by handsome clusters of
> Red-brown, winged fruits in late summer and fall. ...

Often condemned as a weed tree
Because it suckers profusely, but it must
Be praised for its ability to create beauty
And shade under adverse conditions—drought,
Hot winds, and every type of difficult soil.

(Sunset Western Garden Book)

The parallelism between *Ailanthus* and the first Chinese Americans is both obvious and subtle. As befits a complex metaphor, the analogy resides in the combination of several elements. First, as a "Tree-of-Heaven," that is, a "celestial tree," this plant is metaphorically connected with the Chinese, who were dubbed "Celestials" for most of the nineteenth century.[23] Next, these plants, like the diasporic Chinese communities, although China-born ("Native to China"), can be found anywhere in the world (*"All Zones"*), thus reinforcing the adaptability of both the tree and the people, a *topos* of resilience recurring in the last lines of this poem-like description. What stands out most forcibly is the fact that the species is not indigenous to America but has been transplanted from elsewhere, just as Asian Americans had long been considered "transplants from Asia" (Sauling Wong 1993, 9).[24]

Lines 3–5 provide the historical explanation for the presence of *Ailanthus* in the American West, an account that, once more, uses metaphor to link the tree to Chinese Americans: *"Planted a century ago in/ California's gold country where it now/Runs wild."* The fact that *Ailanthus*'s history so closely resembles that of the first Chinese immigrants, who also arrived in Gold Rush California in the mid-nineteenth century, does not escape the attentive reader's notice. There is also a possible suggestion that *Ailanthus* may have arrived in California in the hands and pockets of those first immigrants. The author, in choosing to open the book with the full description of the tree, highlights the negative connotations that accrued to both plant and human species alike. Like *Ailanthus,* these immigrants from China soon came to be perceived as an exotic invader who, like the plant, had begun to "run wild" all over California.[25]

The obvious danger in conflating environmentalist and socio-cultural categories, as in the idea of an "invasive species" as the correlate of a human group, is that we risk falling into a discourse of xenophobic exclusion reminiscent of Yellow Peril ideologies. As Ursula Heise cogently argues, ethnic and biological diversity are incommensurable and do not necessarily follow the same logic (2008a, 400). Irrespective of what racist, xenophobic discourses might imply, human beings can never be classed as an invasive species to be weeded out. Inversely, neither can we apply the positive insights gained from social research, including the need to embrace and celebrate multicultural diversity, to environmental practices.[26]

The final references to *Ailanthus* offer a variation on the leitmotif of adaptability and resilience so often associated with Asian Americans.[27] The last lines emphasize both the aesthetic pleasure and the usefulness of the tree and, by extension, of the Chinese American community: "*Often condemned as a weed tree*," this plant deserves recognition "*for its ability to create Beauty/And shade under adverse conditions.*" Weeds and weed-flowers thus become the privileged trope to explore the resistance and survival strategies deployed by Asian immigrants.[28] However, the figurative use of plants like *Ailanthus* still responds to anthropocentric tenets which, as Sarah McFarland explains, generally reduce the role of "nonhumans" to functioning as mere "metaphors or mirrors for human interests" (2013, 155).

2.2   From Analogy to Human Inlanding

If Shawn Wong finds in the "naturalized" *Ailanthus* a useful trope that contributes to unearthing the hidden history of Chinese America, he finds in the larger American environment an even more powerful tool: the literal bonding of human beings with the land. It may come as a surprise that, despite their different ideological positions at the time, Kingston mobilized the same rhetorical tool in *China Men* a year later. Early on in *Homebase*, the narrator introduces us to his great-grandfather and describes his and his fellow countrymen's hard work building the railroad through the Sierra Nevada mountains. It is at this point that Rainsford imagines his ancestor looking for a quiet place to sleep at night. After a strenuous climb, the great-grandfather approaches a waterfall and rejoices at "the moonlit river at night" (9). He moves closer to the falls, his "skin tast[ing] the air" (9), and when he reaches the top, he stops and starts sweating: "The mist from the crashing falls soaked him and mixed with his sweat" (9–10). The description, like the character himself, slows down: "He rested for a moment, looking down into the river's valley. The water appeared vague, uncertain, it became the sound of moonlight, rather than the sound of water rushing through the valley. The *moonlit mist carved valleys* out of the granite, not the river" (10; emphasis added). The narrative indicates that it is not the thunderous river, as one might expect, but the apparently harmless "mist" that has eroded those American sierras. The careful reader will recall at this point that, a few sentences earlier, the same mist blended with the ancestor's sweat. The metaphorical resonance is evident: the combination of natural elements and human labor has gradually changed the face of the earth. In more ways than one, the author seems to suggest, Chinese immigrants left their trace on those American mountains.

The implication of this scene is that the apparently weak element in the overall schema finally defeats the Goliath forces through "conscious" resilience, be it that of the misty carver of the granite mountains or that of Chinese immigrants surviving in a far-from-welcoming American society. Like the description of the *Ailanthus* tree, this landscape episode may be read as another metaphorical moment. However, I would argue that, on this particular occasion, the scene is not merely metaphorical: it points to the character's closer intimacy with a land that initially seemed hostile to workers who had to "fight" those granite mountains to make way for the Transcontinental Railroad. Granted that analogies drawn from the natural world can be instrumental in unearthing and clarifying the hidden histories and forgotten ancestors; here, however, the author has moved from the metaphorical axis of the tree analogy to a metonymic one. The anonymous workers' metaphorical sweat (toil) has claimed America for them; their literal sweat (liquid), emanating from a physical human body, mixes with the American landscape, claiming contiguity and even, as we shall see, conflation with this Gold Mountain.

Soon after the scene at the falls, Rainsford's great-grandfather concedes that this apparently hostile environment, with its harsh weather, has finally overpowered him. The images in the following excerpt conjure up a kind of sweet surrender, as if the man has let himself be seduced by the land:

> He knew he was stuck here. In Wyoming, the thunderstorms moved in every day. ... The raindrops made the dust rise from the ground, filled his nostrils with the smell of moist earth, he felt the ache in his body rise as the dust rises in the wide meadows. His *giving in to America*, here, was the violence of his soul and he felt it, chased it, and *let it overcome him*. After the rains, the humidity rose and moist air mixed with the dust the rain had raised. It was a *good smell* and he *bared his chest to that air*. (10; emphasis added)

A more somber aspect of the new seductions of the land resurfaces later in the chapter. When his wife comes to join him from China, the great-grandfather seems content, even if they have to live separately for months. One day, while waiting for him in San Francisco, she awakes to a sunlit morning and an intriguing sensation: "this could not be the city, its stench, its noise replaced by this sweet air" (13). She soon realizes that what she is experiencing is her husband's visitation: "this air, this breath, was her husband's voice" (13), trying to contact her from the faraway mountains. The very land, however, gets in the way of this bonding by lulling her to sleep: "The ground was steaming dry, the humus became her soul. ... She breathed deeply. ... She was complete and whole with that one breath. ... She thought of their son, and ... she tried desperately to reach out to wake ..., but the smell of the humus, the moist decaying leaves struck by sunlight and steaming in her dreams was too much, and she was moving too fast into sleep" (13). Here, the earthly embrace becomes deathly, and in the end the woman literally and metaphorically enters the enticing humus of the earth.

This scene in which the woman's body becomes the soil itself and the previous example of juxtaposition and eventual blending of human secretion (sweat) with a nonhuman element (mist) are not the only cases of literal blending.[29] Another significant moment of body-landscape fusion occurs at the end of the novel. For that episode, a more specific term is needed in order to capture not only the fact that the human bodies literally become part the landscape but also the realization that, as a consequence, those natural landscapes become partly human too. I understand these tropes as examples of human *inlanding*, a concept related to Alaimo's trans-corporeality but which goes beyond trans-corporeal contact and exchange. By *inlanding*, I mean the voluntary or involuntary incorporation of human bodies into natural landscapes, especially those construed as "unspoiled." In the process of inlanding, the human/nonhuman dialectics is resolved in body-landscape synthesis, which can be understood as an ecocentric move or, more accurately here, as a narrative strategy that declares one's belonging in a certain territory by being "landscaped" into it.

Inlanding is both similar to and different from the phenomenon of rhetorical personification (*prosopopoeia*) or anthropomorphization of the land. Like classical anthropomorphic strategies, inlanding makes the human form visible in the landscape. The difference lies in the dynamics of empowerment attached to each phenomenon. When we anthropomorphize the land, *tout court*, it is we as human agents that give shape and value to a particular land formation. The inlanding process, in contrast, is neutral enough to encompass both anthropomorphization and the opposite process, whereby it is the land, not people, that endows the inlanded object—in this case a human body—with shape, power, and value.[30] Even though all literary texts are human narratives and articulate human perceptions, the notion of inlanding allows for a meaningful change in the directionality of agency: the human body does not necessarily impose its shape/will on the mountain; it is the mountain that "borrows" it. For Rainsford, "claiming America" is inseparable from becoming part of the land, of being literally in-the-land or in-landed; however, as we shall see, the narrative position in *Homebase* wavers between benign inlanding and a more traditional anthropomorphization by which the narrator claims America in nationalist terms. When the latter prevails, it only serves to reinforce an anthropocentric attitude of possession and dominion that has little to do with ecological values.[31]

Both the potentially freeing process of inlanding and traditional strategies of anthropomorphization involve the mixing of sentient and non-sentient beings, human and nonhuman realms. Such problematization of the dichotomy between human and nonhuman was quite a bold move for an Asian American novelist writing in the 1970s. Even more intriguing was his use of the iconic figure of the "landscaped Indian," as we shall see in the next section.

2.3 *"Yellowing" the Landscaped Indian*

The first contact between Chinese immigrants and Native Americans probably dates back to the 1860s, when a large contingent of Chinese men was hired to work on the Westernmost section of the first Transcontinental Railroad, commissioned to the Central Pacific Railroad company. An evident example of "railway imperialism" (Davis and Wilburn 1991) or "railroad colonialism" (Karuka 2019), the building of the railway pitted railroad workers, foremen, and barons against the indigenous inhabitants of the lands that the railroad was set to cross, desecrate, and modify forever. The colonial railroad project not only denied Native sovereignty over the lands the Central Pacific Railroad chose to appropriate, it "also relied on an imported labor force managed under conditions of racial violence" (Karuka 2019, 2), Chinese immigrants, whose working conditions were often compared with slave labor (83–86, 100) and who were perceived as expendable tools (89, 96).[32] In a surprising maneuver, for the section linking California to Promontory Summit (Utah), the Central Pacific recruited the Paiutes, whose lands they were traversing, as members of the work force, and allowed them to board the trains for which the railroad was being built (Arrington 1969, 8). Although for the most part "interactions between Chinese and Paiutes," according to Manu Karuka, "were reduced to rumors" (2019, 18), a few articles published in 1868 in different San Francisco journals confirm the existence of some contacts between Paiutes and Chinese laborers at the time. A story published in *Alta California* reported conversations between the newly recruited Chinese workers and Paiutes. More interestingly, another article, this time from the *San Francisco Chronicle* (1868), focuses on the progress that the Central Pacific Railroad company was making at the time and describes how men from surrounding indigenous communities had been asked to work side by side with the Chinese, which posed a practical problem for Crocker—the infamous "railroad baron"—and his foremen. In Crocker's own words, the trouble with these people was "that Indians and Chinese were so much alike personally that no human being could tell them apart and, therefore, for fear of paying double wages, he [Crocker] devised the scheme of employing, working and paying them by the wholesale" (quoted in Kraus 1969, 51). Once the railroad was completed, very few Chinese stayed in Utah or neighboring Nevada: while some returned to China, most went back to California, or on to other jobs along the Pacific Coast, a secondary migration that presumably put an end to their brief contact with Native Americans.

After decades of mutual indifference between the two communities, the ethnoracial revival of the 1960s and 1970s fostered interethnic contact and strategic cross-ethnic alliances. Native Americans and Asian Americans now seemed to have a common goal in their quest for social recognition and rights.[33] Even though actual relations between Native Americans and Asian Americans prior to this emergence of minority activism had been marginal, many Asian American writers were influenced by the Native American movement, as well as by the widespread perception of the American Indian's traditional ways as the epitome of an ecological way of life. The popular stereotype of the "mystical environmentalist" Indian, to use Devon Mihesuah's words (1996, 1997, 57),[34] constituted an obvious case of settler colonial discourse about and construction of indigeneity. The simplified image, which would prove particularly useful to Wong, was a recurrent presence in the 1970s, a decade marked by the emergence of environmentalist awareness in the United States.[35] It was not just the writings but also the visual culture of those years, which "depict[ed] Native Americans as paragons of ecological virtue" (Dunaway 2008, 84), that contributed to the image of Indians as "environmental role models" (Warrior and Smith 1996, 279) and ultimately commodified Native Americans. One immensely popular commercial known as the "Crying Indian," first aired in 1971, capitalized on this stereotype. While the ostensible purpose of the commercial was to raise environmental awareness by appealing to the audience's feelings, the actual aim of this advertising campaign was to stimulate a specific type of consumerism; the strategy was highly problematic: not only was the main actor playing the "Crying Indian" an ethnic fraud, as Dina Gilio-Whitaker reminds us (2019, 103), but the commercial appropriated and exploited indigeneity as a "consumable image."[36] In 1974, the equally polemical "Chief Seattle Speech," popularized by Ted Perry's film script, was exhibited in the US Pavilion at the Seattle World's Fair, preluding the symbolic and social impact it would have in the years that followed.[37]

This recurrent, commodified use of the image of American Indians as emblems of environmentalism can be approached as the last manifestation of a centuries-old discourse of settler colonialism heavily indebted to European thinkers. Ensconced in Rousseau's myth of the "noble savage" lay the assumption that the original bond between human beings and the other creatures and elements of nature was somehow miraculously preserved in those "natural men," quintessentially represented in the United States by Native American peoples.

At the same time, Native and non-Native writers and scholars alike have asserted the belief that, generally speaking, the indigenous communities' relationship with the natural environment is privileged insofar as they perceive the land as sacred and nonhumans as "relations."[38] Leslie Marmon Silko embraces that ecological worldview in both her essays and her fiction, arguing that Native American "[v]iewers are as much a part of the landscape as the boulders they stand on. There is no high mesa edge or mountain peak where one can stand and not immediately be part of all that surrounds. Human identity is linked with all the elements of Creation through the clan: you might belong to the Sun Clan or the Lizard Clan or the Corn Clan or the Clay Clan" (1996, 266).[39] Non-Native environmental scholars have also celebrated the biocentric worldview of cultures like those of Native Americans, with their "ontological egalitarianism" and their "attentiveness to place and local processes" (Manes 1996, 25), or of Australian Aboriginal communities, with their belief in human continuity with the land (Plumwood 1993, 102–3, 148).

In an influential book published in 1999, Shepard Krech III explored the discourse surrounding the "Ecological Indian." This popular but, according to Krech, slightly distorted image construed the Native

American in his/her natural surroundings as someone "who understands the systemic consequences of his actions, feels deep sympathy with all living forms, and takes steps to conserve so that earth's harmonies are never imbalanced and resources never in doubt" (1999, 21; see also 213). While Native scholars recognize the powerful connection between environmental concerns and indigenous communities in America, as noted above, they tend to reject the idea of the Ecological Indian emerging in the last decades of the twentieth century. For Dina Gilio-Whitaker, "the new stereotype set an impossibly high standard to which white environmentalists would hold Native people for the next several decades," and, more worryingly, it attracted non-Native activists who, though well-meaning, generally "tried to fit Indigenous worldviews and practices into their own cognitive frameworks" (2019, 103–104). Non-indigenous environmental scholars have likewise questioned some of the assumptions subtending the stereotype of the Ecological Indian.[40] For Sarah Wald, this stereotyping process both denies the individuality and diversity of Native Americans and places them "outside of time and history" (2016, 165–6), fossilizing them in the figure of the "dead Indian" (King 2013, 53–75; cf. Weaver 1997, 18). Greg Garrard, concurring with Joni Adamson (2001, xvii) in her analysis of the literary potential of confronting and contesting the myth, argues for the need to move "away from the notion of the Ecological Indian towards a nuanced appreciation of the complex ecopolitical issues that permeate contemporary Native American culture and literature; from a poetics of authenticity towards a poetics of responsibility" (2004, 139). This nuanced approach to the environmentalism associated with indigenous communities is more visible in recent publications, by both Native and non-Native scholars, which describe the multiplicity of attitudes found among American indigenous peoples today (King 2013; Estes 2019; Gilio-Whitaker 2019, 91–110; cf. Krech 1999; Wald 2016, 165–6).[41]

In the late 1960s and 1970s, before the discourse of the Ecological Indian had come under critical scrutiny, its popularity among both conservationists and racialized minorities was undeniable. It is probably a combination of the incipient transethnic solidarity of those decades with the ascendancy of such an influential icon at the time that accounts for the admiration with which the occupation of Alcatraz, a crucial event in the Native Americans' struggle for their rights, was received by ethnic minorities. In November 1969, the iconic Alcatraz Island off the coast of San Francisco was occupied by a group of Native Americans, who invoked both federal legislation and Indigenous rights when claiming the island.

The literal and symbolic takeover of Alcatraz was a visually effective way of claiming America, one that Asian American activists and writers like Shawn Wong could not but admire. It is precisely at that point, in the winter of 1969–1970, that Wong situates Rainsford's encounter with a significant Native American character. Chapter Five of *Homebase* starts with a reference to Angel Island, another island in San Francisco Bay, which the narrator's grandfather—like many other Chinese immigrants—got to know well, since it functioned as a detention center from 1910 until 1940. Wong connects the two islands as "places of great sadness and pain" (81) for their inhabitants and claims the land not by right of prior occupation nor by virtue of "discovery," as was cleverly argued by the Native American activists in Alcatraz, but by virtue of inhabitation: "My grandfather's island is Angel Island. It was there that he almost died and that makes it his land" (81).

In the following scene, the narrator becomes one of the inhabitants of Alcatraz during the occupation of the island. Shivering with cold, Rainsford strikes up a conversation with an old Native American and, although he eventually dozes off, their dialogue marks a turning point in Rainsford's life. Where the protagonist of Frank Chin's *The Chickencoop Chinaman* (1972) implicitly looks for a viable father figure in an African American boxer and the boxer's father, in *Homebase* Wong places in a Native American both the responsibility of showing Rainsford where to look for a "home base"—starting with Angel Island—and the burden of giving the Chinese "pioneers" a "face": "an Indian man in whom I saw my grandfather showed me my grandfather's face" (81). Although this could well be in reference to the physical resemblance that links the two human groups, due to a (possible) common origin in northern Asia (83, 85),[42] this visible similarity is further complicated when the old man claims that his ancestors are both Navajo and Chinese (82). Beyond the character's literal resemblance and the suggestion of "mestizaje," Wong ultimately invokes the iconic figure of a Native American to show Rainsford the way to his lost roots for two powerful reasons. Firstly, by drawing parallels between the prison on Alcatraz and the Angel Island Immigration Station, the author points to the shared struggle that unites minority groups. Secondly, and more centrally to our purposes here, linking Chinese American ancestors with Native Americans—in arguably problematic ways—endows the former with the native inhabitants' legitimacy when claiming American land.

The author's belief in the need to unite in a shared struggle for rights and visibility is implicit in the narrative prominence of the Alcatraz take-over in a book supposedly about Chinese Americans. This idea is rein-forced by the fact that Rainsford purposely joins a non-Asian American protest, spending the night on the island even after the old man asks him what he is doing there, occupying Alcatraz with a bunch of Native American activists: "This isn't your battle or your land," he tells him (84). Rainsford engages in an earnest and revealing conversation where the old man pushes him to search for his own origins. At the same time, the chap-ter shows how the two struggles, that of Native Americans and that of Asian Americans, are deeply entangled with each other: as Rainsford explains in response to the old man's query, Alcatraz Island (and the larger Bay Area) is his land too.[43]

The second reason identified above, the "borrowing" of indigenous legitimacy, appears to run counter to the Asian American nationalist agenda, since the very foundation of their claiming of America lay in their ancestors' historical contribution to the building of the country, not in their early occupancy of the American land, the basis of the various indig-enous nations' claims of sovereignty, legally fleshed out in treaties and federal law. Here is where an unusual path seems to open in Wong's proj-ect of claiming America. Faced with the fact that the literal presence of Chinese Americans on American soil only dates back to the nineteenth century, Wong resorts to one final image that bridges such a historical gap: Rainsford's father is in-landed, he gets "landscaped" into the very land that Chinese Americans are claiming as rightfully theirs.

At the end of the novel, Wong invokes the Native American in a strat-egy that combines echoes of the "Ecological Indian" with American Indian entitlement to America as natives of the land. For the narrator, the ultimate proof that Chinese Americans belong to/in the land is that they, like the Native American, are literally "landscaped" into it. To quote Bill Brown, Rainsford "completes his archeological mission by rematerializing his ancestry within the landscape" (1998, 941). This "rematerialization" may be interpreted either as the typical anthropomorphizing, personifying trope, or as a case of ecocentric inlanding:

> And today, after 125 years of our life here, I do not want just a home that time allowed me to have. America must give me legends with spirit. I *take* myths to *name* this country's canyons, dry riverbeds, mountains, after my father, grandfather, and great-grandfather. We are old enough to *haunt* this

land like an Indian who laid down to rest and his body became the outline of the horizon. See *his* head reclining, that peak is *his* nose, that cliff *his* chin, and *his* folded arms are summits. (95; emphasis added)

While there are, of course, extralinguistic referents for this evocation—the reclining human body could refer to any of a number of geographic formations known as "the Sleeping Indian" or "the Sleeping Giant"—the whereabouts of this canyon are not as important as the symbolic significance of the trope and its instrumentality in the process of national territorialization, which, once more, seems to reenact and iterate settler colonial discourse.[44] The textual strategies deployed in the description of the landscaped body, especially the choice of vocabulary, are crucial in determining whether this is a classical case of anthropomorphic possession or a more environmentally conscious trope. The lexical choice fluctuates between the ambiguous, mythical *haunting* and the more powerful *take* and *name*, which, together with the possessive *his*, conjure up the idea of human agency and, possibly, dominion. As happens throughout the novel, the authorial intention is never clear in this regard: the description wavers between the older paradigm of ownership and stewardship, and more ecocentric worldviews.

In order to establish which of the two intentions is more prevalent in Wong's novel, we have to consider the way in which the United States of America became territorialized. From a Eurocentric perspective, settler colonies like the United States, unlike European nation-states, went through a process of territorialization in which it was not so much the people but the specific natural environment that formed the national territory. According to Benedict Anderson's formulation, as Rich Heyman reminds us, "the cultural work of imagining community helps forge a 'deep, horizontal comradeship' that is the foundation for national identity" (2010, 328). Whereas in European countries "national territory is assumed to be *coextensive with the people* who form the 'deep, horizontal comradeship,'" (328), in white settler colonies, like the United States, imperialist discourse implied that the land came first, the imagined community later. Needless to say, prior to the arrival of European settlers, entire continents, such as America or Australia, were perceived or construed as *terra nullius* (Plumwood 1993, 163). In the case of the United States, the obsession with pristine, untouched nature as the source of "true Americanness" meant the symbolic erasure of any indigenous presence (Wolfe 2006).[45] Still, indigenous myths soon came back and haunted that "sanitized" territory.

It is in the context of this fallacious colonial discourse of the virginal, "unspoiled," and uninhabited land that we must understand the reference to the "landscaped Indian" in *Homebase*. In the novel Wong consciously supplements his construction of an Asian American "imagined community" with a strategy typical of settler colonialism: he looks at the natural landscapes first and thinks about the people later. The narrator finds a way to claim America by going to the apparent source, the land itself, and there he discovers the reflection of his own ancestors. Once more, Wong borrows—or, if we read this strategy as co-opting colonialist discourse, appropriates—the indigenous entitlement of American Indians, who have left their imprint on the land to the point that the landscape seems to mirror their presence. However, on this occasion, this mirroring is not presented as an ecocentric inlanding, since it is not the land that actively borrows the human shape, but the human viewer that actively imposes his created myth on the mountain: "We are old enough to haunt this land like an Indian who laid down to rest and his body became the outline of the horizon. This is *my father's* canyon. See *his* head reclining! That peak is *his* nose, that cliff *his* chin, and *his* folded arms are summits" (98; emphasis added). At the end of his quest, therefore, Rainsford metaphorically carves his and his ancestors' names in the American geography and literally changes the "face" of the earth, in such a way that flesh and land merge in the novel's closing image. In order to make the parallel with the image of the Native American "haunting" the very land even more evident,[46] the narrator reproduces the words used in the earlier scene, at the beginning of that chapter, simply but meaningfully adding the phrase *my father's*. With such an apparently insignificant change, the nature of the trope has been radically altered. Behind the possessive adjectives, there lurks not only the shadow of settler colonial discourse, but also a proprietorial undercurrent suggesting that this way of claiming of the land is still operating within the anthropocentric, utilitarian framework, whereby the land is an object to be possessed by human beings.[47]

To be fair, not all examples of inlanding that we find in *Homebase* can be so easily dismissed as problematic forms of anthropomorphization. The ecocentric potential of inlanding is implicit in another significant episode that takes place in Yosemite, where Rainsford's parents take him as a child. There, Rainsford's father shows him a tree stump and explains "the rings of growth" that it exhibits: "Like a blind man, he made me run my fingers over each year grain to feel the year of my great-grandfather's birth, my grandfather's birth, his own birth, and my birth. Out of all this I will see dreams, see myself *fixed in place on the land*" (19; emphasis added). This scene is an early instance of the need for natural and topographic inscription, the need to become part of the land that is being claimed, and anticipates the final trope of the landscaped body. In this case, however, it is not ownership that is conjured up in this quotation, as was the case of the sleeping giant/Indian metaphor, but rather fixedness in the land, inlanding, on apparently equal terms. For all that, the centrality of the image of the landscaped "Indian/Chinaman" with which Wong chooses to end the novel relegates this more compelling trope of inlanding to a secondary position.

We can therefore conclude that *Homebase* provides occasional glimpses of a revolutionary use of in landing; yet, by privileging the landscaped Chinese/Native American as its final, central image, the narrative falls back into an anthropocentric paradigm that uses the natural environment for limited purposes. In this final trope, the creative potential of inlanding is diminished and naming becomes a way of claiming the land in the way conquerors and colonial map-makers did: "I name a canyon after my father" (95). Naming not only helps Rainsford deal with "the tragic chaos of his life" (Brown 1998, 940), but, more crucially, it allows him to fix himself "in place" by resorting to familiar proprietorial tactics. Thus, despite its undeniable literary value and its intriguing use of inlanding tropes, *Homebase* seems to be ultimately predicated on anthropocentric utilitarian tenets: the mountains, the canyon, are "things-for-us," they have been appropriated by the narrator as an object or an instrument, albeit for a laudable cause, that of claiming Chinese Americans' rightful place in American history.

This implicit commodification of the natural environment, never entirely problematized in Wong's novel, will reappear in Kingston's *China Men*. In what follows, I will try to tease out the ways in which Kingston's book similarly echoes the proprietorial, colonialist paradigm, at the same time that it hints at a different approach to the natural environment, more akin to empathy than to ownership.

3 MAXINE HONG KINGSTON'S *CHINA MEN*: CLAIMING AMERICA THROUGH LAND EMPATHY[48]

"Across a valley, a chain of men
working on the next mountain,
men like ants changing the face of the world ..."

"In the end," said Tu Fu, "I will carry a hoe."
(Maxine Hong Kingston, *China Men*)

In attempting to trace the environmental aspects of Kingston's *China Men*, especially the characters' ambivalent relationship with the land, we soon realize that there is more to the Chinese immigration story than meets the eye. The immigrant characters in Kingston's book, much like those in Wong's *Homebase*, have been construed as "conquering pioneers" of the new land, a settler colonialist interpretation that appears to be borne out by the text itself (Yang 2010, 77, 80), and is in keeping with the utilitarian view of the natural environment mentioned earlier. I would argue, however, that the characters in *China Men* exhibit either a special closeness to the land or a penchant for cultivating it, or both, and that this bond problematizes the traditional reading of the text. At one point in the narrative, one of the Chinese immigrants asks himself: "Who really owns the land? … The man who farms it with his sweat and piss or the man whose name is on the paper" (Kingston 1980, 416). In other words, who is an American: a person with citizenship papers, or one who has toiled, sweated, and ached on/for that land?[49] As we saw at the beginning of this chapter, both history and fiction attest to the fact that Asian Americans, like other "pioneers" before them, literally changed the shape of the American land(scape) by building railroads, planting seeds, and harvesting fields. This is precisely the subtext in *China Men*, and, as such, it has been studied and read as contributing to cultural-nationalist efforts to claim America, visibility, and a sense of belonging for Asian Americans.

However, this ethnic nationalist agenda must be questioned and revisited from an ecocritical perspective. In particular, I believe that Kingston's book offers critics that possibility by exploring the tensions that Aldo Leopold situates in the dichotomies of "man the conqueror *versus* man the biotic citizen; … land the slave and servant *versus* land the collective organism" (1949, 223). This ecocritical reading of *China Men* subverts the canonical appraisal of the text as claiming the land in either "conquering" or "servile" terms. As I will try to illustrate, narrative strategies such as Kingston's daring choice of focalization emphasize the characters' shared feeling and connectedness with nonhuman biozens, what can be called "land empathy."

3.1   Land Ethics and Land Empathy

The Asian American cultural-nationalist discourse that "claimed America" in the 1970s implicitly or explicitly approached the land as a commodified object that could be bought and sold. The land was, above all, property to which one was (or was not) entitled.[50] Viewed in such terms, the land itself had no agency or rights,[51] but was passively owned by an active subject, a citizen with land-owning rights. Asian American nationalists sought to be recognized as figuratively *belonging within* the American land while also struggling to attain property rights, that is, to have that land *belong to* them. However, when such rights were finally gained by birth or legal amendment, the subject/object dialectic did not alter substantially. Asian Americans were simply allowed to enter into the category of subjects who could possess, own, and control the object: the land. Important though the recognition of civil and property rights was for Asian Americans, it is no less true that their relationship with the land remained basically unaltered. A profound affective bond between the land and human beings was non-existent, or at best it was pushed to a secondary position. (Human) property rights displaced (nonhuman) land rights and obscured the fact that social ethics should be supplemented by an ethics of ecology that situates human interaction with the land beyond the utilitarian, anthropocentric paradigm.

Aldo Leopold was one of the first environmental activists to advocate such a revolution in our understanding of the world, when he called for a new, extended ethical imperative in *A Sand County Almanac* (1949). In a chapter entitled "The Land Ethic," Leopold sketched the ideal progression that this ethical development should take: from individual, to collective, to ecological ethics—although his own ethical hierarchy can be rather problematic at times.[52] In later decades, deep ecologists and environmental philosophers became equally critical of the Western instrumental approach to the natural environment and its nonhuman inhabitants. Lovelock's Gaia hypothesis, which became popular in the 1970s, helped further a more holistic understanding of the planet, stressing our interdependence with other organisms and rejecting what Val Plumwood would later define as the "hyperseparation" of human beings and nature. In *Feminism and the Mastery of Nature*, Plumwood accuses traditional Western philosophy, not only of promoting such hyperseparation but also of viewing nature in exclusively utilitarian terms, "as a means to the self-contained ends of human beings" (1993, 147). In our Western worldview, the land has been parceled out and its pieces perceived as mere "real estate, readily interchangeable as equivalent means to the end of human satisfaction" (147).[53] As an alternative to this exclusively instrumental understanding of nature, Plumwood proposes a new ethical attitude: "that human beings be concerned with [nonhuman] others for their own sake and that one's ends make ineliminable reference to the ends of others" (151). Inspired by these environmental philosophers, most notably Leopold and his extended notion of ethics, Greta Gaard and Patrick Murphy similarly denounce the human tendency to see nonhuman elements in terms of self-interest: as "things-for-us," instead of "things-in-themselves" (1998, 5–6). Against this utilitarian view of the nonhuman members of the biotic community, ecocritics like Gaard and Murphy urged us to consider them "things-in-themselves," thus endowing them with intrinsic worth, a value more typically associated with human beings in many Western cultural and religious traditions.[54]

If Leopold's "land ethic" was still a utopian desire in the 1940s, it remains a marginal conviction today. Although much has been done in the way of environmental education and consciousness-raising, and we human beings are starting to realize, voluntarily or involuntarily, our dependence *on* the planet, our interdependence *with* the planet is a different matter. While most of us are well aware that we are part of a social, human community, it is less clear whether we all enlarge "the boundaries of the community to include soils, waters, plants, and animals," what Leopold summarizes in the concept of "the land" (1949, 204).[55] Leopold's insight that we cannot broaden those collective boundaries without learning about and learning to love that new, extended biotic community leads us to the concept of "land empathy."[56] For a land ethic to exist, therefore, there must be "an ecological conscience," which "in turn reflects a conviction of individual responsibility for the health of the land" (221). After all, as Leopold suggests, developing a land ethic involves "an intellectual as well an emotional process" (225). In other words, not only do we have to grasp our biotic interdependence *intellectually*, we also need to take it in at an *emotional* level: we must *learn to feel* an integral part of that biotic community.[57] The next section traces the journey into land empathy by some of the characters in *China Men*, as they learn to feel land empathy by attributing not only human features but also "interiority" to the nonhuman Other.

3.2   Homelands: Homelessness and Belonging

In one section of *China Men*, entitled "The Laws," Kingston quotes a section of the 1868 Burlingame Treaty, regulating immigration between China and America. The fifth article of this international agreement invokes an interesting new right, that of "curiosity":

> The United States of America and the Emperor of China cordially recognize the inherent and inalienable right of man to change his home and allegiance, and also the mutual advantage of the free migration and emigration of their citizens and subjects respectively for the one country to the other *for the purposes of curiosity* ... (373; emphasis added)[58]

That apparent openness and mobility, as we now know, was not without its limits; yet many people succumbed to their "curiosity," especially in Southern China. As the narrator in *China Men* puts it, Cantonese villagers, like ocean people, had itchy feet, and, in their restlessness, they traveled to the Sandalwood Mountains (Hawai'i) and the Gold Mountain (mainland United States): "the ocean and hunger and some other urge made Cantonese people explorers and American" (306). The narrator's father, BaBa, was no exception. When leaving, Chinese men vowed to their wives and mothers that they were mere "tourists," sojourners, birds of passage; they promised "to come back, not settle in America with new wives" (257). In the end, many of the men who left for America eventually settled in the new land and started to think of themselves as (at least partly) American. They constituted what came to be known as the "bachelor society," including old-timers like the narrator's grandfathers and more recent arrivals like her father. Nonetheless, despite the many ancestors that had lived in America, until the "Brother in Vietnam" finally secures a Q Clearance, "certifying that the family was really American, not precariously American but super American" (257), the status of the narrator's family in the United States remains doubtful. This is textualized by the narrative wavering between "legal" and "illegal" versions of the narrator's ancestors' lives, most notably her father's. In the legal version of BaBa's life, "The American Father," the narrator emphasizes his power to "mak[e] places belong to him": small, specific places marked by "father smells," but also larger places like America (463). Similarly, in one of the "illegal" versions of her father's coming to the United States, the narrator tells how BaBa hid in a box as a stowaway on a ship bound for America, "coming to claim the Gold mountain, *his own country*" (266; emphasis added). In another version of the story, as we saw in Chap. 2, BaBa arrives in America by ship and endures a long period of anguished waiting in the Immigration Station building, such a long time that he starts to doubt the very existence of a city that the fog prevents him from seeing. Despite the lingering nature of that doubt, San Francisco turns out not to be a mirage after all. Fellow detainees at Angel Island not only assure BaBa of the solidity of

America, but also of their belonging in that territory even before setting foot on it: "Let me land. I want to come *home*" (270). Significantly enough, when her "legal father" finally passes the Immigration interview, he feels he has "won America," thus confirming the initial conception of the land as "conquerable" (274), a perception heavily invested with settler colonialist presumptions.

The cultural project of claiming America is also played out in varying ways in the stories of other male ancestors in *China Men*. Significantly, these ancestors not only claim American territory in figurative terms but also reclaim the land in literal, technical terms, either by turning it into agricultural land (Bak Goong's land reclamation in Hawai'i) or by making room for the new railway road (Ah Goong in the Sierra Nevada mountains).[59] "Crazy" Ah Goong is probably the most contradictory of the ancestor characters. His relationship with the land that he helps reshape, as he digs and blows up mountains, is a mixture of awe and tenderness, as we shall see later in this chapter. In his search for a sense of belonging, he seems more astute than his fellow countrymen, who still cling to the traditional idea of homeland. For most of the nineteenth century and the beginning of the twentieth century, the bones of Chinese immigrants who died in America were sent back to China, the "genuine homeland." However, when these deaths took place in the dangerous Sierra Nevada mountains, more often than not the corpses could not be rescued. The workers' bodies were eventually left there to nurture the American land (much like the Chinese wife who surrendered to the humus in Wong's *Homebase*). Even when the corpses were found, sending them back to China was not the only option. At one point in Kingston's *China Men*, there is a tragic accident while the men are building the railroad and, when the bodies of the Chinese laborers are finally recovered, there is talk of burial ceremonies in the new, American land. This suggestion is met by suspicious resistance from most of the sojourners and open reluctance from those who, before dying, are still able to mumble a few last words: "'Don't leave me frozen under the snow. Send my body home. [...] Aiya. To be buried here, nowhere'" (357–58). To this, however, visionary pioneers like Ah Goong answer in a positive vein: "But this is somewhere. [...] This is the Gold Mountain. *We're marking the land* now. The track sections are numbered, and your family will know where we leave you" (358; emphasis added). On this occasion, the literal marking of the railway track is enhanced by a figurative reading highly redolent of animal territory-marking, a "natural" action that also contains connotations of possession.

Ah Goong's feeling of belonging in the new land, his attempt to claim America as home, reappears in Bak Goong, the great-grandfather of the Sandalwood Mountains, their name for Hawai'i.[60] There, Bak Goong works long days at the sugar cane plantations and, on his "day offu" (323), he joins his fellow countrymen and goes to Honolulu: "They dressed for town in their black silk suits and yellow straw boaters with the black band and forked streamers; their braids hung between the swallowtails. 'I'm going to town,' Bak Goong sang and sang *town* as if Honolulu were *his own village*" (324; emphasis in the original). Bak Goong ultimately goes back to China and thus fulfills the promise made to his wife before leaving (321). However, in one of his feverish nightmares he dreams that he returns to China penniless and physically wasted, and is scolded by his wife, who reminds him of his new belonging and allegiance: "Moneyless and bodiless, you better go back to the Sandalwood Mountains. Go back and pick up your money and your body. *Go back where you belong.* Go now" (334; emphasis added). It is there, in those islands, that the narrator's great-grandfather seems to belong now.

Another intriguing character in *China Men* is Kau Goong, the Great Uncle who looked like a giant to the young narrator. After many years in California, Kau Goong receives news from his wife, who wants him to go back to China, where she awaits him. Despite Great Aunt's insistence, the old man finally opts to stay in America: "'California. This is my home. I belong here.' He turned and, looking at us, roared, '*We* belong here'" (407). When Kau Goong dies, the other Chinese old-timers who attend the funeral tell everybody how he chose to stay in his new home: "He was a Gold Mountain Man. They said 'Gold Mountain' a lot, 'Gum Sahn' many times. 'Long time Califoon,' they said in English" (409). To make up for the blatant omission of their contribution to the United States from the official history of the country, these "China men" and their relatives have produced an alternative oral narrative, one in which they inscribe themselves not only as Gold Mountain pioneers but as "genuine" Americans.[61] Old Chinese immigrants, the narrator's grandfathers, and her extended family all join in the rites of claiming that land as "ancestral ground":

> When their descendants came across the country to visit us, we took them
> to the place where two of our four grandfathers had had their house, stable,
> and garden. ... They took pictures with a delayed-shutter camera, everyone
> standing together where the house had been. The relatives kept saying,
> "This is *the ancestral ground*," their eyes filling with tears over a vacant lot
> in Stockton. (391; emphasis added)

In response to the erasure of the Chinese pioneers' presence from
official history, Kingston reimagines her ancestors as silently claiming America by claiming the land. This belated recognition arrives in
the guise of the sacralization of "minor" spaces that synecdochically
stand for larger ones. Thus, the "vacant lot in Stockton," California,
stands for the supposedly vacant land which Kingston's ancestors have
worked in and/or cultivated: a foreign space that they turn into their
new home(land), into "the ancestral ground" where their memory can
finally rest.[62]

Although some of these ancestor characters, in particular Bak Goong,
Ah Goong, and Bak Sook Goong, eventually go back to China, their
sojourn in America has helped forge a deep bond with the places they
inhabited for years. Bak Goong, as we have just seen, embraces Honolulu
as "his own village" and the Sandalwood Mountains as his "home." Yet
the same Bak Goong often feels homesick and longs to go back to China.
Like the king in the prince-kitten story who cannot afford for his subjects
to discover the "unnatural aberration" of an heir with cat's ears, but at the
same time desperately wants to let the secret out,[63] Bak Goong cannot
keep his secret any longer, so he transgresses all the silence norms of the
plantation and shouts it out (335–36). The homesick Chinese sojourners
dig "a wide hole" and all slump down, lying there "with their faces over
the edge of the hole and their legs like wheel spokes," shouting into the
big hole, just as in the prince-kitten story that precedes the scene. In this
case what Bak Goong, Bak Sook Goong, and the other pioneers have
done, in symbolic terms, is to dig "an ear into the world" in order to tell
"the earth their secrets":

> "I want home," Bak Goong yelled, pressed the soil, and smelling the earth.
> "I want my home," the men yelled together. "I want home. Home. Home.
> Home. Home."
> Talked out, they buried their words, planted them. (337)

A close reading of the text itself reveals how the men's original intentions are thwarted in several ways. Firstly, the earth seems to contradict their desires, because the echo of their shouts, bouncing back and around in the hole, reminds them that "home" may be right where they are. By having their words "buried," "planted" in that alien land, the workers symbolically leave part of themselves behind, like Ah Goong in the Sierras (351–52). Secondly, even in those moments of nostalgia when the great-grandfather concedes that he wishes he were back in China, he also acknowledges his foundational attachment to the new land, combining the questionable adjective employed by the settler colonial phrase "Founding Fathers" with the equally polemical appropriation of indigenous (Hawai'ian) ancestral rights: "That wasn't a custom [in that island]. ... We made it up. We can make up customs because we're the *founding ancestors of this place*" (337). Finally, just as in the traditional story, where the earth was unable to hold the king's secret for very long, here in Hawai'i the landscape would also tell of these immigrants' lives and longings: "Soon the new green shoots would rise, and when in two years the cane grew gold tassels, what stories the wind would tell" (338).[64] The significance of this episode is reinforced, as we shall see next, when echoed in the narrator's own personal memories of her visit to the Sandalwood Islands.

3.3 Claiming America Through Land Empathy

While there is critical consensus that the different stories in *China Men* are linked by the leitmotif of claiming America, little attention has been paid to the manner in which the nationalist agenda both determined and was determined by these men's personal relationship with the land as railroad-construction and agricultural workers. I believe that Kingston uses her narrative to chronicle the awakening of a sense of land empathy, based on her characters' first-hand experience of their new environment. Two of the characters that are essential for this project are Bak Goong, the great-grandfather of the Sandalwood Mountains, and Ah Goong, the "crazy" Chinese grandfather of the Sierra Nevada mountains.

In China, Ah Goong, like other farmers in a predominantly agricultural society, had "plowed fields hour after hour alone, inching along between earth and sky," but in America he would have to carve the very land and shape it into something new (226). From the very outset of the narrative, the character of Ah Goong displays a strangely intimate connection with the earth. The unusual story of his arrival in America sets the tone for his life. While still in his village, Ah Goong had heard the planet move, "the gears on its axis snap," and, by following a faraway sound resembling the fireworks during a New Year holiday, he had finally reached "the blasting in the Sierras" of the American continent (346). In historical terms, this means that the narrator's grandfather, like many other Chinese sojourners in America, arrived in the country sometime around 1863, just as the transcontinental railroad was being built.

Once in America, Ah Goong and the other Chinese laborers, "men like ants changing the face of the world" (351), start to reconfigure the land in a literal sense, by blasting their way through mountains and clearing valleys. When Ah Goong first arrives, the railroad employers tell him to clear the land of trees. In order to do so, he engages in some "bloody" battles with trees that not only seem to resist being cut or uprooted but, more importantly, take on human or animal characteristics:

> He axed for almost a day. ... The tree swayed and slowly dived to earth, creaking and *screeching like a green animal*. He was so awed, he forgot what he was supposed to yell. Hardly any branches broke; the tree sprang, bounced, pushed at the ground *with its arms*. The *limbs* did not wilt and fold; they were a small forest, which he chopped. The trunk lay like *a long red torso; sap ran from its cuts like crying blind eyes*. At last it stopped fighting. (346–47; emphasis added)

Already in this early task, the character-focalizer Ah Goong endows trees with sentient characteristics typically associated with human beings or animals (arms, torso, "crying blind eyes," "screeching like a green animal"). The tree he has to cut is not a passive object, but is explicitly given agency, to the point that its fierce struggle overwhelms the new lumberjack, who is "awed" and, to quote Leopold's words, becomes "humbly aware that with each stroke [of the axe] he is writing his signature on the face of the land" (1949, 68).

Next, the ant-like workers try to work their way through the imposing granite mountains and their initial failure emphasizes their insignificance: "After tunneling into granite for about three years, Ah Goong understood the immovability of the earth. Men change, men die, weather changes, but a mountain is the same as permanence and time. This mountain would have taken no new shape for centuries, then thousand centuries, the world a still, still place, time unmoving" (354). As Hayashi rightly observes of this passage, hard work provides Ah Goong with "a perspective on the environment ... reminiscent of the way Aldo Leopold comes to appreciate the natural world in his famous piece 'Thinking Like a Mountain,' considering the detached and long-range perspective of a mountain" (2007, 69). In contrast to Leopold, Hayashi adds, the view of this mountain is far from idyllic; instead, it represents a conscious effort to honor and remember "the human toll of American labor history, not just an acknowledgment of its natural history. The land is full of stories, full of ghosts—not a once virginal place outside of culture" (69–70). The whole book, indeed, bears witness to the fact that even those territories construed as pristine wilderness are also laden with "an extraordinary amount of human history," to quote Raymond Williams (1980, 67).

The work of Chinese immigrants was not only hard, therefore, but also highly dangerous. Many workers risked their lives, hanging precariously from cliffs in order to do their job. When these laborers fell and died, their bodies joined the rocks and debris below, thus literally becoming part of the land. Only when dynamite replaced gunpowder in the construction of the railroad was the granite finally "subdued," but the cost was high for those workers: "Human bodies skipped through the air like puppets and made Ah Goong laugh crazily as if the arms and legs would come together again. *The smell of burned flesh remained in rocks*" (356; emphasis added). Not only are the mountains made part of the pioneers through their sweat and vice versa, as happened in *Homebase*, but these men also "impregnate" the land with their blood, urine, and semen. This striking image of Ah Goong's unconventional way of "claiming" the land bears quoting in full:

> The mountainface reshaped, they drove supports for a bridge. Since hammering was less dangerous than the blowing up, the men played a little; they rode the baskets swooping in wide arcs; they twisted the ropes and let them unwind like tops. "Look at me," said Ah Goong, pulled open his pants, and pissed overboard, the wind scattering the drops. "I'm a waterfall," he said. He had seen *a part of himself* hurtling. On rare windless days he watched his piss fall in a *continuous* stream from himself almost to the bottom of the valley. (351–52; emphasis added)

Although it has been argued that hard work, particularly agricultural labor, has historically rendered "creative, sensitive response to the land difficult, if not impossible" (Smith 2007, 11), the quotation above demonstrates that Ah Goong, playful as ever, finds ways to engage creatively with the land to the extent that he emerges as a metaphorical continuation of the surrounding environment, with his peculiar "waterfall" completing the picture. Much the same may be said of another bodily fluid that literally connects him with the land:

> One beautiful day, dangling in the sun above a new valley, not the desire to urinate but sexual desire clutched him so hard he bent over in the basket. He curled up, overcome by beauty and fear, which shot to his penis. He tried to rub himself calm. Suddenly he stood up tall and squirted out into space. "*I am fucking the world*," he said. *The world's vagina was big, big as the sky, big as a valley*. He grew a habit: whenever he was lowered in the basket, his blood rushed to his penis, and he fucked the world. (352; emphasis added)

This surprising scene invites many and multifarious interpretations. It has been read as indicative of Ah Goong's literal and metaphorical impotence, a metonymic reference to the symbolic castration of the early Chinese American community (Hayashi 2007, 66). It has also been interpreted as "a triumph for the body" if not for the whole person, "a moment of transcendence" for an otherwise alienated worker (Yang 2010, 77). The passage could also be construed, from an ecofeminist perspective, as a metaphorical "rape" of the land. Following the analogy that Annette Kolodny establishes in *The Lay of the Land* (1975), the penetration of feminized nature by a male figure would signify yet another form of sexual exploitation.[65] However, the fact that the fear-induced scene is told in a celebratory, even triumphant tone points to another interpretation: the narrative shows Ah Goong joyfully bonding not only with that specific place, the Sierra Nevada mountains and valleys, but also with the planet. Hayashi has rightly warned critics not to forget that, "unlike other shapers of the American landscape," Asian Americans "often had little say in what their 'contribution' would be" (2007, 65); nevertheless, they did have enough freedom to imagine or reformulate the manner in which they perceived that contribution. While heeding Smith's and Hayashi's caveats against idealizing agricultural or railroad workers, there remains an urgent need to rescue those instances when perceptual and imaginative agency overcomes social and material constraints.

Just as Ah Goong and his fellow immigrants reconfigure the land in continental America, in Hawai'i Bak Goong helps to "shape" the islands by clearing the land for the sugar cane plantation. Chinese workers like Bak Goong cut down thickets and move rocks, while the Hawai'ians, more reluctant to interfere with their ancestral land, "quit rather than help pull the boulders out of the earth" (320).[66] When the back-breaking work of clearing the land finally comes to an end, the feat is described following the (problematic) masculinist script of traditional fairy tales: "One day, like a knight rescuing a princess, Bak Goong broke clear through the thicket," and the bullocks "yanked out" the last stumps (320). Gradually, however, land reclamation is replaced with land empathy. Like the trees in the American Sierras, which fell "with a tearing of veins and muscles" (347) when cut down or uprooted with gunpowder, the soil of the Sandalwood Mountains metamorphoses into a suffering creature, bleeding like a human or nonhuman animal. Thus, in a passage that demonstrates the workers' consciousness of the land's sacrifice, the narrator explains how the Chinese "pioneers" have the sad honor of being "the first human beings to dig into this part of the island and *see the meat and bones of the red earth*. After rain, *the mud ran like blood*" (320; emphasis added).[67] Contrary to Smith's contention that hard labor tended to instill a "sense of alienation from the land" in workers (2007, 93), this moment signals the emergence of an incipient land empathy among those Chinese immigrants. Having chosen these male ancestors as the main focalizers—if not the narrative voice—of *China Men*, Kingston links up the humanizing metaphors with the workers' interiority. The choice of words reveals the intense empathy that these "China men" feel with and for the land. In feeling with and for the trees, the animals, and the rocks, they embody the ethical broadening advocated by Leopold, "the extension of the social conscience from people to land" (1949, 209). These men learn to value the other members of the biotic community, not just as "things-for-us" but as "things-in-themselves."

While hard work in the natural environment is the privileged context in which the ancestors forge their special bond with the land, there are two other important paths in *China Men* that lead to land empathy: gardening and visions. One way to respond creatively to the land and establish an affective bond with it is to plant a garden. The conspicuous presence of gardening in Asian American history and literature may be read as a strategic attempt by Asian Americans to circumvent their disenfranchisement in terms of land property and other forms of ethnic discrimination.[68] Kingston not only shows Bak Goong's engagement in gardening, but emphasizes that he does it for pleasure, even after a lifetime of agricultural work: "For recreation, because he was a farmer and as antidote for the sameness of the cane, he planted a garden near the huts. ... He even grew flowers, for which there was no edible use whatsoever" (323). Bak Goong is not alone in his non-utilitarian love of gardening. Whether in the cane field or in their own private gardens, the slow growth of plants filled many Chinese pioneers with a sense of fulfillment that they might not have encountered elsewhere.[69] However, as the Chinese American population became increasingly urban, the risk of believing that "breakfast comes from the grocery," to quote Leopold (1949, 6), became more evident. No longer farmers, Chinese immigrants started to live in American cities— first New York and then Stockton and San Francisco—and "neglect attending the big public celebrations" as well as "the planting and harvest days," which "made no difference in New York" (289). BaBa seems to be an exception to the rule. Known for his "green fingers," he is defined as "a man who enjoys plants and the weather" (222). Significantly, when BaBa finally has a house of his own in America, in Stockton, the first thing he does is plant a vegetable garden.

A rather different way in which the narrator's ancestors establish a spe-cial bond with the land in *China Men* is through visions that allow the characters to see beyond the surface of reality. In one of BaBa's anxiety-induced visions, for instance, the frightening seascape he cannot see becomes a landscape he can delight in. Trapped inside the box where he is hiding as a stowaway on a ship bound for America, BaBa starts to feel "the ocean's variety—the peaked waves that must have looked like pines; the rolling waves, round like shrubs, the occasional icy mountain; and for stretches, lulling grasslands" (263). These soothing visions of nature are combined with more powerful epiphanies, like Bak Goong's opium-induced vision in which he learns to "embrace opposing thoughts at the same moment. He loved the strangers around him as much as he loved his family. He closed his eyes and saw islands in the sea or planets in space or lakes he could dive into or observe" (311). The umbilical cord linking all members of the biotic community becomes literal in Bak Goong's epiph-any: he sees the futility of human-made bridges, "when there is already an amazing gold electric ring connecting every living being as surely as if we held hands, flippers and paws, feelers and wings. Though he was leaving his good wife and his village, they were connected to him by a gold net or a light; it shimmered when the people *and other creatures* moved about" (311, emphasis added). Visions like these are reminiscent of both Buddhist mysticism and deep ecology's holistic worldview.[70]

The most captivating natural epiphany, however, has little to do with holistic or Buddhist visions. It is strategically placed in the first-person

section where the narrator describes her visit to the Hawai'ian island originally known as Mokoli'i and later, in an (Asian) settler colonialist move,
renamed as Chinaman's Hat (Ho'omanawanui 2008, 134). According to
Kingston, this name derives from the conical shape of the island, reminiscent of the hats worn by Chinese workers in the nineteenth century and
construed by the narrator as a "tribute to the pioneers" (2008, 305). Here
on the island, the process of claiming the land reaches a poignant climax,
as the narrator becomes part of that pioneering ritual at last. When she first
arrives in Hawai'i with the explicit desire to discover and "claim" the
Chinese legacy in those islands, she seems unable to distinguish the pioneers' voice (303). It is only when she plunges into the ocean and swims
toward the island, when she literally enters the space she means to claim,
that she manages to hear their song:

> a howling like wolves, like singing, came rising out of the island. "Birds,"
> somebody said. "The wind," said someone else. But the air was still, and the
> high, clear sound wound through the trees. It continued until we departed.
> It was, I know it, the island, the voice of the island singing, the sirens
> Odysseus heard.
>   The Navy continues to bomb Kaho'olawe and the Army blasts *the green
> skin* of the red mountains of O'ahu. But *the land sings*. We heard something.
> (305; emphasis added)

Significantly, the narrative voice and the focalizer are one and the same
here: the autodiegetic narrator. At this particular moment, she finally hears
the land singing: she bears witness to her ancestors' stories "by listening in
the cane" (305), in those islands where Ah Goong had "planted" his
words. The whole scene at Chinaman Hat constitutes, as noted by Patricia
Linton, "a complex and powerful metaphor that demonstrates how people appropriate the land, making it so completely their own that the land
itself tells their story" (1994, 42). I would argue, however, that there is
another, underlying narrative that supplements the paradigm of claiming
America. By letting the land itself sing/speak as a subject, nature is imagined no longer as a passive object, but rather as an active agent in a cultural-
nationalist project which, while originally anthropocentric, has broadened
to include a new ecological sensibility.[71]

Kingston's *China Men* claims America in subtly different ways from
those traditionally acknowledged by critics. By textualizing the visible
imprint of bones, blood, and other bodily fluids that Chinese Americans

left on the very territory they were claiming as theirs, the narrator, true to her cultural-nationalist agenda, makes a powerful statement: America belongs to these "pioneers" because they have physically become the territory they are claiming. In this regard, Kingston's strategies are similar to those wielded by Wong in *Homebase*, since both, wittingly or unwittingly, echo the settler colonial model. On the other hand, Kingston's book contains signs of a new relationship with the land, one that has moved beyond the anthropocentric paradigm of domination and possession. In these instances, the narrative combines the awareness of nature's agency and of its "sentient" features, on the one hand, with the characters' incipient sense of "land empathy," on the other. Enacting Leopold's theory that "a land ethic changes the role of *Homo sapiens* from conqueror of the land-community to plain member and citizen of it," which entails both "respect for his[/her] fellow-members, and also respect for the community as such" (1949, 204), the characters in *China Men* eventually move away from proprietorial understandings of land and toward an egalitarian bond with the natural environment. Contrary to what might be expected, considering the alienating conditions around them, the Chinese "pioneers" learn to feel with and for the once hostile land. Not only does Kingston's text offer us a glimpse of nature's feelings, but it also allows us to hear its voice: *natura sentiens, natura loquens*.[72] Through this new dialogue between human and nonhuman, the old process of "claiming America" is thus dismantled and built anew.

4   Agency and Voice: Can (Subaltern) Nature Speak?

China Men offers several instances in which the land seems to speak or sing to human characters, but can we really speak of a *natura loquens*? To what extent is nature allowed to have agency and voice in our (human) discourse? And, as a "subaltern," can nature speak or do we speak for it/her/him? As many scholars have observed, the question of whether their "object(s)" of study can represent themselves is what separates ecocriticism from other politically engaged schools, such as postcolonial, ethnic, and gender studies. Even though environmental criticism usually "incorporates questions of social justice," as Rigby reminds us, "it nonetheless differs from other forms of political critique" insofar as it fights for the rights of an entity considered "unable to speak for itself" (2015, 137).

Kari Weil voices the concern that, for animals, it seems hardly possible to escape the human representations that have circumscribed them and have "justified their use and abuse by humans" for centuries, because, while nonhuman animals may communicate, their language differs from "the languages that the academy recognizes as necessary for such self-representation" (2010, 1–2; 2012, 4). Similarly, McFarland has noted the need to move beyond anthropocentric models, at the same time conceding that to speculate about the "self-world" of animals is to do so "in terms of human experience and using human language; the empathy necessary is somewhat illusory because there is no other way to speculate" (2013, 155). This is even more obvious in living organisms from outside the animal kingdom, such as plants, and of apparently inert nature, such as rocks. Mountains may be thinking, but they cannot tell us their thoughts, which makes it even more imperative to learn to think like a mountain, to echo Leopold's famous dictum.

Nevertheless, even if we try to envision nature as an active agent, the question remains of whether our current anthropocentric and logocentric discourse can accommodate a speaking nature. In "Nature and Silence," Christopher Manes criticizes philosophers who try to approach environmental problems from a humanist, rationalist perspective. For Manes, it is the discourse of rationalist humanism itself, with the fiction of "Man" at its center, that needs to be destabilized, because it precludes the very possibility of a non-silent nature. In contrast to animistic societies, in contemporary Western culture, "Nature *is* silent … in the sense that the status of being a speaking subject is jealously guarded as an exclusively human prerogative" (1996, 15; emphasis in the original). For Manes, human hübris and lack of "ecological humility" lie at the root of the current ecological crisis: "it is within this vast, eerie silence that surrounds our garrulous human subjectivity that an ethics of exploitation regarding nature has taken shape and flourished" (16). The message implicit in his critique is that we have to discard our anthropocentric beliefs in order to embrace a more salutary environmental ethics.[73]

Some of the characters in Kingston's *China Men* appear to have learned "the language of ecological humility" advocated by Manes (17). They no longer approach the natural environment as a passive, silent object: instead, as we have seen, they feel the land suffering with them, imagine the wind telling stories, and manage to hear the land sing. After all, we are embodied "in a landscape that 'speaks' to us through sensory experience" (Goodbody 2014, 66). If we refuse to listen to that sensory language, the fault lies with us, for "the view that nature is silent might well say more about our refusal to hear than about nature's inability to communicate" (Rigby 2015, 138). The main characters in *China Men* are testament to the truth of this statement: they learn to listen to what nonhuman nature has to tell them. When BaBa lies hidden in a ship bound for America, he hears the ocean and is convinced that it is trying to talk to him: "The sea invented words too. He heard a new language, which might have been English, the water's many tongues speaking and speaking. Though he could not make out words, the whispers sounded personal, intimate talking him over, sometimes disapproving, sometimes in praise of his bravery" (264). Similarly, when the narrator swims toward the island, she hears the land sing. These are some of the ways devised by Kingston to allow the environment to speak in her narrative.[74] Therefore, it may be argued that, in *China Men*, Kingston has debunked the prevailing discourse, one which "has produced a certain kind of human subject that only speaks soliloquies in a world of irrational silences" (Manes 1996, 25). Her characters have managed to go beyond an appreciation of natural beauty and enter into an incipient dialogue with the nonhuman environment. If we accept Manes's thesis that there is a clear "link between listening to the nonhuman world (i.e., treating it as a silenced subject) and reversing the environmentally destructive practices modern society pursues" (16), characters like BaBa and the attentive narrator herself may be seen as having made the transition from ecological awareness to an ethics of environmental care. Allowing (subaltern) nature to speak and, more importantly, having humans listen to the natural environment in an unprejudiced manner may be the first step toward embracing both ecological and humanist concerns.[75]

NOTES

1. As described in the second chapter, the Gold Mountain was both a physical site, corresponding to the North American region to which the discovery of gold in the mid- and late nineteenth century drew countless immigrants, and a mental site in the collective imagination of nineteenth- and early twentieth-century Chinese people. According to the narrator in Kingston's *China Men*, the Gold Mountain was nothing but a Chinese "invention" (302). That "invention," however, remains metonymically associated with the emergence of Asian America, even though California, the area signified by the Gold Mountain, is by no means the only part of the country that is home to Asian Americans today.

2. In Masumoto's own words, "I'll consider myself a better farmer when I have a clearer sense of history about a place, when I understand the knowledge of a farm's hills and the sweat and blood left behind. Until then I'm just managing a piece of dirt and probably still foolishly believing I rule the earth" (1995, 113). Here, and elsewhere in *Epitaph for a Peach*, Masumoto takes a rather skeptical view as regards the possibility of "claiming the land" (1995, 111), at least literally, as a farmer.

3. *China Men* received the National Book Critics Circle Award for Non-Fiction in 1981. *Homebase* was awarded the Pacific Northwest Booksellers Award for Excellence in Writing (1980) and the 15th Annual Washington State Governor's Writers Day Award (1980).

4. The phrase is used by Kingston herself in Timothy Pfaff's "Talk with Mrs. Kingston" (1980, 1). According to Shirley Lim, the term "'claim' brings to mind a claim as a piece of land especially for mining purposes, recalling that the first groups of Chinese American came to the United States to work in the goldfields of California. Kingston's statement asserts an identity right and an empirical history of mining and land claims that became gradually disallowed through state and federal laws barring Chinese from land ownership and U.S. citizenship" (2006, 299).

5. In the last two decades, ecocriticism has departed more and more from these earlier regionalist and localist efforts, in what has been variously described as the "transnational turn" or "global shift." Recent ecocritical studies, which we will examine in more detail in Chap. 6, have announced that important shift in environmental criticism. Ursula Heise, a proponent of the transnational turn, pits this new eco-cosmopolitanism against earlier environmental models such as Leopold's land ethic or bioregionalism. She claims that, rather than clinging to such old discourses, twenty-first-century ecocritics should be concerned with developing "environmental allegiances that reach beyond the local and the national" (2008b, 21). Nevertheless, much environmental activism still relies on the appeal to the autochthonous and the local, and recent ecocritical publications prove that regional and local impetus has by no means disappeared from environmental scholarship. Traditional ecocritics such as Patrick Murphy continue to emphasize the relevance of "localism" and bioregionalism for environmental criticism: "ecocriticism should remain ... localist, rather than global, in its grounding orientation" (2009, 1). Environmental philosophers such as Val Plumwood note that, despite its dwindling effectiveness, bioregionalism has an undeniable value (1993, 186). Some examples of recent ecocritical scholarship that privilege the local and find the global shift problematic are Murphy's *Ecocritical Explorations in Literary and Cultural Studies* (2009), Richard Evanoff's *Bioregionalism and Global Ethics* (2011), and *The Bioregional Imagination* (2012), edited by Tom Lynch, Cheryll Glotfelty, and Karla Armbruster.

6. The indigenous peoples' struggle to recover their full sovereignty should be understood within its proper framework, one that views the United States as perpetuating the British colonial project in the American continent and engaging in new imperialist ventures like the occupation of Hawai'i. I am aware that describing the history of Asian Americans' "civil rights struggles as one of nation building" (Fujikane and Okamura 2008, 2) may risk obscuring the Asian settlers' role (and often their complicity) in the colonial project that deprived indigenous peoples of their land. "North America," as Manu Karuka reminds us in *Empire's Tracks*, is "a space of imperialism, an international space of hundreds of colonized Indigenous nations, rather than of settler 'nations'" (2019, 176; see 168). Therefore, in the discussion that follows, I have tried to heed these scholars' advice, noting while not endorsing the settler colonialist premises often reflected in the narratives under scrutiny. I thank the anonymous reader for their advice regarding this issue and for providing me with an updated bibliography that helped me engage in some last-minute revisions of this chapter.

7. This term is commonly found in the discourse of the Yellow Peril explored in the third chapter. Skeldon perceptively notes that this "contrast between sojourners and settlers, often in the context of comparing Asian, mainly Chinese, movements with European migration, can be overdrawn. Many Europeans were also sojourners and returned relatively quickly to their home countries" (1996).

8. It goes without saying that not all white citizens were landowners, but no law denied them the right to own land, unlike in the case of Asian Americans.

9. As Bill Brown highlights, the very name of the protagonist in *Homebase*, Rainsford, points to a literal and figurative displacement: the name "marks a disjuncture," because the town he is named after "has disappeared from the map" (1998, 938).

10. Kingston emphasizes this point in the chapter entitled "The Laws": "Though the Chinese were filling and leveeing the San Joaquin Delta for thirteen cents a square yard, building the richest agricultural land in the world, they were prohibited from owning land or real estate" (374).

11. Ecocritical appraisals of African American literature both confirm and refute Smith's thesis. In his 1987 study of the African American novel, Bernard Bell identified a specific Afro-American type of pastoral where one could see the "implicit contrast between country and city life"; the rural-urban dichotomy seems to have haunted African American literary criticism since then. As Sonya Posmentier puts it, most scholarly accounts of black modernism have tended to react against the pastoral movement and consciously "ignore or minimize the presence of the natural world in black

literature altogether" (2012, 275–76). Daniel J. Martin reads Richard Wright's texts and Billy Holiday's "Strange Fruit" as anti-pastoral narratives where the natural environment of the plantation is tainted by its association with racist violence, most notably lynching. Thus, Martin problematizes the American pastoral ideal, for "the pastoral leaves out many people who are not white" (2007, 107; see Outka 2008). Some recent studies, however, tend to offer a more optimistic reading of African Americans' relationship with the natural environment or else combine the tainted pastoral approach with alternative interpretations. In *Shades of Green: Visions of Nature in the Literature of American Slavery, 1770–1860*, Ian Frederick Finseth focuses on antebellum writers, among them Frederick Douglass, whose book, *My Bondage and My Freedom* (1855), is read as linking "not only the southern fields with the blood of his enslaved people," as we would expect, but also "the systematic violation of African Americans with the systemic pollution of the natural world" (2009, 281). In *Black on Earth: African American Ecoliterary Traditions*, Kimberly N. Ruffin notes the "burden-beauty paradox" faced by African Americans in their relationship with nature, since they have had to deal with environmental racism in the form of degraded environments, while at the same time benefiting from the solace that a beautiful environment can bring (2010, 2–3).

12. For Karuka, the connections between slavery and indentured work are obvious, since "the fear of Chinese labor ... revolved around an equation of Chinese labor with slave labor, and its impact on white labor" (2019, 100).

13. Gifford also analyzes what he calls the post-pastoral mode, a term he uses not so much in a chronological as in a conceptual sense. The critic identifies this as "the essential paradox of the pastoral," since the "retreat to a place apparently without the anxieties of the town, or the court, or the present, actually delivers insights into the culture from which it originates" (Gifford 1999, 82). The discourse of (post)pastoralism will be explored in more detail in later chapters.

14. The contrast between country and city, Williams notes, usually "depends ... on just the suppression of work in the countryside, and of the property relations through which this work is organized" (1973, 46). Already in the Renaissance, the real particularities, the "living tensions," were "excised" until only the "selected images" survived, as Williams puts it, "not in a living but in an enamelled world" (18).

15. As Williams observes in "Ideas of Nature" (1980), nature is inextricable from human interaction and, more specifically, from human labor. Both Williams and White emphasize how hard it is to find "untouched" territories in our natural environment: "We seek the purity of our absence, but everywhere we find our own fingerprints" (White 1995, 173).

16. The same conclusion is repeated in Caroline Yang's "Indispensable Labor: The Worker As a Category of Critique In *China Men*" (2010). For an excellent introduction to our need to hide "the true cost, both to subordinate humans and to the earth, of our production processes and consumption habits," see Rigby (2015, 122–23).

17. A preliminary version of the following analysis was presented at the 32nd International AEDEAN conference, held in Palma de Mallorca (Spain) in 2008, and a summary was published in the conference proceedings.

18. In 1972, Frank Chin, Jeffrey Paul Chan, Lawson Fusao Inada, and Shawn Wong published their first collective piece on Asian American literature in the *Bulletin of Concerned Asian Scholars*, an essay later reprinted in the first edition of the *Aiieeeee!* anthology (1974).

19. The new prefaces to the first anthology's Mentor edition and to its sequel, *The Big Aiieeeee!* (1991), together with other collective essays, repeated the anthology's original call for Orientalizing tendencies be curbed and stereotypes debunked (see "*Aiieeeee!* Revisited," xxxvi).

20. For an analysis of gender issues in *Homebase*, specifically of the white "dream bride" in the novel, see Sakurai (1993).

21. The novel thus blends "roots" and "routes" before that combination came to be heavily theorized in diaspora and transnational studies. Within Asian American literary criticism, *Homebase* has been studied as a paradigmatic example of the trope of mobility in Asian American literature (Sau-ling Wong 1993).

22. Of the strategies Rainsford uses to anchor himself in the American land, one of the most conspicuous is the evocation of place names: the narrator confesses that he tries to "[r]oot down [his] life into the names of places" (23), as seen in the echoes of both his forename and his last name, here reimagined as a demonym: "Chan is short for California" (1–2). For Rainsford, the "chronicling of [his] life should be given the name of a place" (95).

23. According to the *Merriam Webster Dictionary*, the demonym derives from an "old name for China": "*Celestial* Empire," itself the translation of *Tianchao*.

24. Even Edith Eaton, the first author to publicly acknowledge her Asian ancestry as Sui Sin Far, was marketed by her editor Charles Fletcher Lummis as "a Chinawoman transplanted" to America.

25. It was ironic, to say the least, that "after two centuries of European-introduced diseases had swept across the Americas and decimated native peoples, Americans viewed Asian immigrants as threats to the land's well-being" (Hayashi 2007, 66).

26. Heise argues that cultural solutions cannot be easily transferred to ecological concerns. In "Ecocriticism and the Transnational Turn," Heise warns us to beware of narratives that take for granted the argument "that cultural and

biological diversity refer to analogous structures in social and ecological systems," for that "assumption often leads to arguments and forms of narrative logic that make little ecological sense" (2008a, 400). In this case, though trying to keep the ethnoracial other away is not ethically acceptable and does not make social sense, protecting autochthonous species from invasive ones makes a lot of environmental sense. See Heise 2008a (383–87) and Heise 2008b (28–49). For a different approach to the similarities between social and environmental fields, see Gamber (2012, 12–13).

27. Other natural metaphors commonly associated with resilient, adaptable Asian Americans are bamboo, used by Kingston in *The Woman Warrior*, and water, employed by Bruce Lee in his popular "Be Water, My Friend."

28. As we shall see in the fifth chapter, the use of this trope will reappear in the narratives dealing with the Japanese American internment.

29. This apparent exchange between the materiality of human bodies and "the more-than-human world" can be read in terms of trans-corporeality, as defined by Stacy Alaimo (2010, 2), a concept to which I will return in later chapters.

30. Brown interprets the anthropomorphizing strategy differently, as "the final act in a drama of projective incarnation where artifacts and the artifactual serve as the mode of keeping the past proximate, and of keeping it distant, of turning the imagoes of the psychic life into physical objects" (1998, 941).

31. Contrast the predatory attitudes of settler colonialism with the more organic bond with the land espoused by some indigenous nations. In recent years, Native American scholars have worked with the concept of either internal colonialism (Weaver 1997; cf. Wald 1987) or settler colonialism (Estes 2019; Gilio-Whitaker 2019). For an overview of twentieth-century settler colonialism, see Elkins and Pedersen (2005).

32. As Karuka cogently puts it, "Chinese labor was an in instrument, not a subject, of colonialism" (2019, 81); still, it was an integral part of the project of "continental imperialism" in its westward expansion.

33. What came to be known as the Asian American Movement emerged in "the wake of the civil rights movement in the late 1960s" (Cheung 1997, 1) and had much in common with the Chicano, Black, and Red Power movements of the time. See Warrior and Smith's *Like a Hurricane* for a detailed account of the American Indian movement in the late 1960s and early 1970s, from the occupation of Alcatraz to Wounded Knee. For a description and analysis of recent Native American protests and occupations, such as the peaceful resistance and struggle against the Dakota Access Pipeline (DAPL), see Estes (2019) and Gilio-Whitaker (2019). Both scholars frame their discussion within the larger history of indigenous activism in North America.

34. In a similar fashion, but from a non-Native perspective, Buell uses ironic circumlocutions such as "the model minority sage of green wisdom" or the "eco-sensitive indigene" (2005, 23, 24).

35. According to Adam Rome (2003), the seeds had already been planted in the previous decade, but it is generally agreed that it was the first Earth Day, celebrated on April 22, 1970, that launched a decade of environmentalism, including the first environmentally themed World's Fair in Spokane, Washington, in 1974.

36. I thank the anonymous readers of this manuscript for emphasizing the colonialist nature of this advertising strategy: a "greenwashing initiative meant to benefit bottlers," instead of responding to "Native self-articulations," in the readers' own words.

37. See Rudolf Kaiser's comprehensive study of the different versions of the speech/letter by Chief Seattle (1987, 497–502). For an account of the fabrication (and appropriation) of the Ecological Indian, see Feest (2003, 10–13), Krech (1999, 214), Garrard (2004, 120–25), Harkin and Lewis's 2007 collection, and Gilio-Whitaker (2019, 103–104, 138). For a brief description of the most prominent Native American stereotypes, see Mihesuah (1996); for a highly readable debunking of myths surrounding indigenous people in North America, see King (2013).

38. According to Nick Estes, this is precisely the term employed by the Lakota to refer to nonhuman animals (2019, 15).

39. Recent examples include Lee Schweninger's "To Name Is To Claim," in *Coming Into Contact* (2007) and *Listening to the Land: Native American Literary Responses to the Landscape* (2008), and "Landscape as Narrative, Narrative as Landscape" (2010), by Theresa Smith and Jill Fiore. See also Joni Adamson's *American Indian Literature, Environmental Justice, and Ecocriticism: The Middle Place* (2001), and Donelle N. Dreese's *Ecocriticism: Creating Self and Place in Environmental and American Indian Literatures* (2002).

40. In his "Native Americans and the Environment," David Lewis explains that Native Americans have never been "ecologists," properly speaking, although they were "careful students of their functional environments," and, accordingly they "developed an elaborate land ethic based on long-term experience, tied to a cosmological view of the world with all its animate and inanimate, natural and supernatural inhabitants as an interrelated whole" (1995). For Christian Feest (2003), it is a mistake to automatically credit Native Americans with "environmental virtue."

41. At the end of *The Ecological Indian*, Krech describes the different attitudes in among North American Indian communities, ranging from conservationism to utilitarian views of the land (1999, 213–28), to the point of antagonizing environmentalists from time to time (222, 225).

42. As we shall see in Chap. 5, this physical resemblance is relevant in Perry Miyake's *21st Century Manzanar*. It also appears in a seminal Native American novel written by Leslie Marmon Silko, *Ceremony* (1986), where the resemblance prompts the protagonist to blend the faces of Japanese soldiers and those of fellow Native Americans.

43. It bears reminding that the history and situation of indigenous and immigrant populations in North America are widely different. Such incommensurability, however, should not prevent us from noticing the analogies drawn by non-Native American authors, like Wong, even though such symbolic "borrowing" of indigenous sovereignty, as we shall see next, may be read as replicating or co-opting the claims of white settler colonialism.

44. Among the landscapes that have also been imagined as "sleeping Indians" in the Americas, "the Sleeping Giant" in Thunder Bay (Canada) has been described as "a formation of mesas and sills that juts out on Lake Superior and forms the body of water that is Thunder Bay. When viewed from the City, this remarkable peninsula resembles a reclining giant. ... An Ojibway legend identifies the Sleeping Giant as Nanabijou, who was turned to stone when the secret location of a rich silver mine was disclosed" http://www. thunderbay.ca/Visiting/Sleeping_Giant.htm.

   For a slightly different version of the story, see Smith and Fiore (2010, 70). For other Sleeping Giant mountains, see http://www.mnmuseumofthems. org/Faces/FacesGiant.html and http://astro.wsu.edu/worthey/astro/ html/im-indian-heads/indian.html.

45. Anthropologist Patrick Wolfe explores how settler colonialist discourse encouraged the *terra nullius* vision, consciously erasing the presence of indigenous communities from the land in order to legitimize European imperialist ventures. For a description of the ethnic cleansing involved in the creation of US National Parks, especially the "erasure" of Native Americans, see Kantor (2007) and Oatman-Stanford (2018).

46. The textual implication that real indigenous people, having been erased from the North American landscape, simply become ghosts of the past haunting the land, dangerously echoes the famous settler trope of the Vanishing Indian. Popularized as much by the nineteenth-century classic, James Fenimore Cooper's *The Last of the Mohicans,* as by the twentieth-century Hollywood (see Gilio-Whitaker 2019, 59), the image still holds sway today, albeit in the more radical guises of the "Indian as Corpse" (Weaver 1997, 18) or the Dead Indian (King 2013, 53–75).

47. For a different reading of the inscription of human bodies in the landscape, see Brock (2012).

48. The following analysis was first presented at a conference on Maxine Hong Kingston, held at the University of Mulhouse in March 2011. The paper

was later expanded and published as a chapter (Simal 2013) in Sämi Ludwig and Nicoleta Alexoae-Zagni's book celebrating Kingston's legacy. In a later publication, Andrea Aebersold similarly argues that, in *China Men*, "Kingston conceptualizes the workers and the land as an inseparable entity" (2015, 19). Likewise, in her recent study of *The Nature of California* (2016), Sarah Wald also notes the link between "claiming the land" and claiming citizenship: "Claiming to belong (at least figuratively) to the natural environment of the United States is one way to claim national belonging" (2016, 9).

49. The narrator highlights this question when she points out that "[e]ven if Ah Goong had not spent half his gold on Citizenship Papers, he was *an American for having built the railroad*" (366). Crazy Ah Goong finally becomes a pariah and has to be brought "home," back to China, or he dies in San Francisco, we never know for sure: "The family called him Fleaman. They did not understand his accomplishments as an American ancestor, a holding, homing ancestor of this place" (372). Note, once more, the problematic "borrowing" of indigenous legitimacy.

50. It is important to recall that the "origins of property rights in the United States are rooted in racial domination" (Harris 1993, 1716), since African Americans became commodified, that is, turned into property, through slavery. Likewise, Native Americans were deprived of their lands, which were taken over by white Europeans: "Only white possession and occupation of land was validated and therefore privileged as a basis for property rights" (278). Taken together, these processes led to the creation of what Cheryl Harris terms "whiteness as property" (278). For a discussion of the related concept of "shareholder whiteness," see Karuka 2019 (149–167).

51. See Gilio-Whitaker 2019 (154–157) for a description of the new legal notion of Rights of Nature (RON).

52. The most striking example of Leopold's unethical subtext can be found early in the book, when his racist/xenophobic prejudices, in this case against Chinese people, become quite obvious: "The erasure of a human subspecies [sic] is largely painless—to us—if we know little enough about it. A dead Chinaman is of little import to us whose awareness of things Chinese is bounded by an occasional dish of chow mein. We grieve only for what we know" (1949, 48).

53. In *The Question Concerning Technology*, originally published as "Die Frage nach der Technik" (1949), Martin Heidegger explored the instrumentality brought about by modern technology: "The earth now reveals itself as a coal mining district, the soil as a mineral deposit" (14). Nature, according to this instrumentalist view, is merely *Bestand* or "standing reserve" (17). However, rather than criticize this model and look for alternatives, Heidegger limits himself to describing modern technology and how it shapes our world(view).

54. Linda Hogan describes that utilitarian approach to nonhuman nature, which "sees life, other lives, as containers for our own uses and not as containers in a greater, holier sense," as "far-hearted" (2001, 312), and contrasts it with the generally more respectful indigenous worldviews. Needless to say, not all Western traditions adhere to such pragmatic or exploitative agendas; within Christianity, for instance, the Franciscan view of nature is at loggerheads with utilitarianism.

55. This idea has recently been taken up by Allewaert and Ziser when they argue in favor of an enlarged type of "commons" or "space … in which human and environmental needs are addressed together"; a collectivity "including not only human beings and institutions but the seas in which nuclear and other wastes circulate, atmospheres that are getting hotter, corals and other sea creatures that grow in the places humans leave vacant, and turbulences, whether those of divers dropping art into the seas or of the larger stochastic forces that have the power to remake the scene entirely" (2012, 241).

56. Although Leopold does not use the phrase explicitly, it may be surmised that this is what he was aiming at when he declared the following: "We can be ethical only in relation to something we can see, feel, understand, love, or otherwise have faith in" (214). Leopold's assertion that contact/contiguity makes love is highly problematic from an environmentalist point of view. If we focus only on the local, what becomes of our concern for the larger planet? Not only that, but, as Heise rightly observes, most place-oriented scholars and bioregionalists are misguided in taking for granted "that sociocultural, ethical, and affective allegiances arise spontaneously and 'naturally' at the local level, whereas any attachments to larger entities such as the nation or beyond require complex processes of mediation" (2008b, 34).

57. In this context it is surprising that Plumwood should accuse Leopold of ignoring emotional attachment to the nonhuman as a reason for broadening the ethical imperative, what she terms "moral extensionism" (1993, 170).

58. Kingston has recently suggested it might be time to bring this "right" back: "I feel that we should all copy this paragraph [from the Burlingame Treaty], and cut it out and laminate it and keep it with our passport. If we get to a checkpoint or if you forget your passport, you just bring this out, show it to them: 'Look! I have the right to my curiosity. … The borders are open. Travel is free.' I think that as writers and scholars we should always remember that we have the right to travel freely" (2013, 26).

59. These actions illustrate the immigrants' problematic involvement in the settler colonialist project. In *Empire's Tracks*, Karuka reproduces the initial positive attitudes of San Francisco journalists, who saw the Chinese workers as "reclaimers of the soil" and "remakers" of the landscape, thus crucial elements in the settler colonialist project in California (2019, 95).

60. See Fujikane and Okamura (2008) for a study of Asian settler colonialism in Hawai'i.

61. For an analysis of Kingston's *China Men* as an intervention in and rewriting of official history, see Sledge (1980) and Cheung (1990).

62. This claim to "ancestrality" is, once more, highly reminiscent of indigenous sovereignty rights. When associated with the idea that working (in) the land gave immigrants the right to claim it, at least in a symbolic way, it seems to replicate the arguments deployed by colonialist discourse. We need to remember that the accusation often leveled at indigenous peoples, the fact that they did not cultivate or use the land in the same manner as the European settlers, became the colonial excuse to divest Natives of their rights of sovereignty over those lands. For a discussion of how the discourse of settler colonialism effectively disenfranchised Native Americans, see Weaver (1997), King (2013), and Estes (2019).

63. The story is a variation of the classical metamorphosis of King Midas narrated by Ovid in Book XI of his *Metamorphoses*. Where Midas is given donkey ears by the gods, here a prince is born with feline appendages.

64. Alternatively, the scene may be read as a plea for understanding that the entire planet is your real home, since the echo in the hole, which keeps repeating "home, home, home" can also be interpreted to be the voice of the earth itself, speaking and shouting back that we all belong together. Caroline Yang has read this passage in different terms. For her, "the text highlights Bak Goong and the other workers' practice of freedom in acts of resistance, such as 'let[ting] out scolds disguised as coughs' (104), throwing a Grand Ball to 'prove their civility' (109), and digging a hole into which they can shout their desires and complaints against their working conditions (117). The last act in particular signals a turning point for Bak Goong in how he is treated, as he 'talked and sang at his work, and did not get sent to the punishment fields' after the 'shout party'" (2010, 75). For Yang, the "worker's subversive use of song can be read with African American plantation songs in mind" (75).

65. Where Ah Goong symbolically impregnates the Sierra Nevada mountains, Bak Goong also leaves some seed in the Sandalwood Mountains: after sowing the field Bak Goong "pats the pregnant earth," in a more pastoral and paternalistic fashion than "crazy" Ah Goong. These examples confirm the fact that the natural environment, the land, has been "symbolically coded as female—an arena of potential domination analogous to the female body" (Buell 2005, 109). Kolodny emphasizes the ambivalent result of feminizing the land, when she "draws attention to the conflict between phallic and foetal attitudes towards the feminised landscape, whereby the impulse to penetrate and master the country as a whole has oscillated uneasily with a desire to preserve certain places perceived at once as 'virginal' and 'maternal'" (Rigby 2015, 136).

66. This would reinforce Ho'omanawanui's contention that, in contrast with Native Hawai'ians, Asian Americans (including those who settled in the islands) continued to see "Hawai'i as a commodified resource, not as an

ancestor" (2008, 123). For a critical analysis of the book's portrayal, or rather non-portrayal, of indigenous Hawai'ians as well as women workers, see Yang (2010, 74–76). For a general study of the prolonged ostracism of indigenous peoples and their views in Hawai'i, see Fujikane and Okamura (2008).

67. Aebersold reads this episode where the mud is likened to blood as linking environmental and human pain, since both nature and the workers "are sacrificed for economic gain" (2015, 19).

68. As Smith puts it, "aesthetic and spiritual interaction with nature, as through gardening" can be interpreted as a way of countering ethnic/racial discrimination (2007, 95). More specifically, these gardens provided a safe haven for persecuted Chinese immigrants during the "Driving Out" which followed the completion of the railroad, an episode also recollected in the novel (367). As we shall see in Chap. 5, Japanese American "internment" gardens are also endowed with historical and symbolic connotations.

69. When the reporters interview one of the Chinese American pioneers at his birthday party, they ask him what has given him the most joy in his one-hundred and six years, to which he answers: "What I like best is to work in a cane field when the young green plants are just growing up" (539).

70. We find other examples of these visions in *China Men*: when the characters spend the night camping or sleeping under the open sky, they are overcome by conflicting feelings of utter loneliness and sublime immensity (234, 348–49, 465).

71. This reading would contradict Ho'omanawanui's argument that Asian settlers, and Asian Americans in general, see Hawai'i as merely "a picturesque setting for people centered stories not as a character" (2008, 134).

72. The Latin term *natura loquens* was suggested by the title of the 5th EASLCE Conference "Natura Loquens," held at the Universidad de La Laguna, Tenerife (Spain), in June 2012.

73. In recent years, as we shall see in some detail in Chap. 6, ecocritical theory has tried to answer these and similar questions. Materialist ecocritics, for instance, not only argue for the agency of the material environment, but also read materiality as "storied matter," matter that tells stories and engages in "conversations" (Iovino and Oppermann 2014, 1–4).

74. The same strategy reappears in *Through the Arc of the Rain Forest* (1990), where Yamashita endows the metaphorical planet with the narrative voice, as we shall see in the Chap. 6.

75. This need to listen has been endorsed by recent nature writers. "Science made us arrogant," notes Teresa Jordan, convincing us "that the earth should bend to our will"; however, learning how "we have damaged the systems that sustain us" can ultimately have salutary, humbling effects: "Perhaps, just perhaps, it can pave the path of our redemption. We have tried to tell the earth what to do; maybe now we can learn how to *listen*" (2001, 317; emphasis added).

REFERENCES

Adamson, Joni. 2001. *American Indian Literature, Environmental Justice, and Ecocriticism: The Middle Place.* Tucson: University of Arizona Press.

Aebersold, Andrea. 2015. Environmental Narratives of American Identity: Landscape and Belonging in Maxine Hong Kingston's *China Men* and Milton Murayama's *All I Asking For Is My Body.* In *Asian American Literature and the Environment,* ed. Lorna Fitzsimmons, Youngsuk Chae, and Bella Adams, 13–29. New York: Routledge.

Alaimo, Stacy. 2010. *Bodily Natures: Science, Environment, and the Material Self.* Bloomington: Indiana University Press.

Allewaert, M., and Michael Ziser. 2012. Preface: Under Water. *American Literature* 84 (2): 233–241. https://doi.org/10.1215/00029831-1587332.

Arrington, Leonard J. 1969. The Transcontinental Railroad and the Development of the West. *Utah Historical Quarterly* 37 (1): 3–15.

Bell, Bernard W. 1987. *The Afro-American Novel and Its Tradition.* Amherst: University of Massachusetts Press.

Brock, Richard. 2012. Body/Landscape/Art: Ekphrasis and the North in Jane Urquhart's *The Underpainter. Canadian Literature* 212: 11–32.

Brown, Bill. 1998. How to Do Things with Things (A Toy Story). *Critical Inquiry* 24 (4): 935–964.

Buell, Lawrence. 1995. *The Environmental Imagination: Thoreau, Nature Writing, and the Formation of American Culture.* Cambridge, MA: Harvard University Press.

———. 2005. *The Future of Environmental Criticism: Environmental Crisis and Literary Imagination.* Oxford: Blackwell.

Chan, Sucheng. 1991. *Asian Americans: An Interpretive History.* Boston: Twayne.

Cheung, King-kok. 1990. The Woman Warrior Versus the Chinaman Pacific: Must a Chinese American Critic Choose between Feminism and Heroism? In *Conflicts in Feminism,* ed. Marianne Hirsch and Evelyn Fox Keller, 234–251. New York: Routledge.

———, ed. 1997. *An Interethnic Companion to Asian American Literature.* Cambridge: Cambridge University Press.

———. 2016. *Chinese American Literature Without Borders: Gender, Genre, and Form.* New York: Palgrave Macmillan.

Chin, Frank, ed. 1991. *The Big Aiiieeeee: An Anthology of Chinese American and Japanese American Literature.* New York: Meridian.

Chin, Frank, Jeffery Paul Chan, Lawson Fusao Inada, and Shawn Wong. 1972. Aiiieeeee! An Introduction to Asian American Writing. *Bulletin of Concerned Asian Scholars* (Fall 1972). Reprint, 1974. In *Aiiieeeee: An Anthology of Asian American Writers,* ed. Chin et al. Washington, DC: Howard University Press. Reprint, 1991, xxi–xlviii. New York: Mentor Books.

Davis, Clarence B., and Kenneth E. Wilburn Jr., eds. 1991. *Railway Imperialism*. New York: Greenwood Press.

Dreese, Donelle N. 2002. *Ecocriticism: Creating Self and Place in Environmental and American Indian Literatures*. New York: Peter Lang.

Dunaway, Finis. 2008. Gas Masks, Pogo, and the Ecological Indian: Earth Day and the Visual Politics of American Environmentalism. *American Quarterly* 60 (1): 67–99. https://doi.org/10.1353/aq.2008.0008.

Elkins, Caroline, and Susan Pedersen. 2005. *Settler Colonialism in the Twentieth Century: Projects, Practices, Legacies*. New York: Routledge.

Estes, Nick. 2019. *Our History Is the Future: Standing Rock Versus the Dakota Access Pipeline, and the Long Tradition of Indigenous Resistance*. London: Verso.

Evanoff, Richard. 2011. *Bioregionalism and Global Ethics: A Transactional Approach to Achieving Ecological Sustainability, Social Justice, and Human Well-Being*. New York: Routledge.

Feest, Christian. 2003. The Greening of the Red Man: "Indians", Native Americans, and Nature. In *"Nature's Nation" Revisited. American Concepts of Nature from Wonder to Ecological Crisis*, ed. Hans Bak and Walter W. Holbling, 9–29. Amsterdam: VU University Press.

Finseth, Ian Frederick. 2009. *Shades of Green: Visions of Nature in the Literature of American Slavery, 1770–1860*. Athens: University of Georgia Press.

Fujikane, Candace, and Jonathan Y. Okamura, eds. 2008. *Asian Settler Colonialism: From Local Governance to the Habits of Everyday Life*. Honolulu: University of Hawai'i Press.

Gaard, Greta, and Patrick Murphy, eds. 1998. *Ecofeminist Literary Criticism: Theory, Interpretation, Pedagogy*. Chicago: University of Illinois Press.

Gamber, John B. 2012. *Positive Pollutions and Cultural Toxins: Waste and Contamination in Contemporary U.S. Ethnic Literatures*. Lincoln: University of Nebraska Press.

Garrard, Greg. 2004. *Ecocriticism*. London: Routledge.

Gifford, Terry. 1999. *Pastoral*. London: Routledge.

Gilio-Whitaker, Dina. 2019. *As Long as Grass Grows: The Indigenous Fight for Environmental Justice, from Colonization to Standing Rock*. Boston: Beacon Press.

Goodbody, Axel. 2014. Ecocritical Theory: Romantic Roots and Impulses from Twentieth-Century European Thinkers. In *Cambridge Companion to Literature and the Environment*, ed. Louise Westling, 61–74. Cambridge: Cambridge University Press.

Harkin, Michael E., and David R. Lewis, eds. 2007. *Native Americans and the Environment: Perspectives on the Ecological Indian*. Lincoln: University of Nebraska Press.

Harris, Cheryl I. 1993. Whiteness as Property. *Harvard Law Review* 106 (8): 1710–1791.

Hayashi, Robert T. 2007. Beyond Walden Pond: Asian American Literature and the Limits of Ecocriticism. In *Coming into Contact: Explorations in Ecocritical Theory and Practice*, eds. Annie Merrill Ingram, Ian Marshall, Daniel J. Philippon, and Adam W. Sweeting, 58–75. Athens: University of Georgia Press.

Heidegger, Martin. (1949) 1977. The Question Concerning Technology. In *The Question Concerning Technology and Other Essays*, 3–35. New York: Garland.

Heise, Ursula K. 2008a. Ecocriticism and the Transnational Turn in American Studies. *American Literary History* 20 (1-2): 381–404.

———. 2008b. *Sense of Place and Sense of Planet: The Environmental Imagination of the Global*. New York: Oxford University Press.

Heyman, Rich. 2010. Locating the Mississippi: Landscape, Nature, and National Territoriality at the Mississippi Headwaters. *American Quarterly* 62 (2): 303–333.

Ho'omanawanui, Ku'ualoha. 2008. This Land Is Your Land, This Land Was My Land. In *Asian Settler Colonialism*, ed. Candace Fujikane and Jonathan Y. Okamura, 116–154. Honolulu: University of Hawai'i Press.

Hogan, Linda. 2001. What Holds the Water, What Holds the Light. In *Getting Over the Color Green: Contemporary Environmental Literature of the Southwest*, ed. Scott Slovic, 310–313. Tucson: University of Arizona Press.

Iovino, Serenella, and Serpil Oppermann. 2014. Introduction: Stories Come to Matter. In *Material Ecocriticism*, ed. Serenella Iovino and Serpil Oppermann, 1–20. Bloomington: Indiana University Press.

Jordan, Teresa. 2001. From Playing God on the Lawns of the Lord. In *Getting Over the Color Green: Contemporary Environmental Literature of the Southwest*, ed. Scott Slovic, 314–317. Tucson: University of Arizona Press.

Kaiser, Rudolf. 1987. Chief Seattle's Speech(es): American Origins and European Reception. In *Recovering the Word: Essays on Native American Literature*, ed. Brian Swann and Arnold Krupat, 497–536. Berkeley: University of California Press.

Kantor, Isaac. 2007. Ethnic Cleansing and America's Creation of National Parks. *Public Land and Resources Law Review* 28: 41–64.

Karuka, Manu. 2019. *Empire's Tracks: Indigenous Nations, Chinese Workers, and the Transcontinental Railroad*. Oakland: University of California Press.

Kim, Elaine. 1982. *Asian American Literature: An Introduction to the Writings and Their Social Context*. Philadelphia: Temple University Press.

King, Thomas. 2013. *The Inconvenient Indian: A Curious Account of Native People in North America*. Minneapolis: University of Minnesota Press.

Kingston, Maxine Hong. (1976) 2005. *The Woman Warrior*. New York: Everyman's Library, Knopf.

———. (1980) 2005. *China Men*. New York: Everyman's Library, Knopf.

————. 2013. Opening Speech. In *On the Legacy of Maxine Hong Kingston: The Mulhouse Book*, ed. Sämi Ludwig and Nicoleta Alexoae-Zagni, 21–29. Berlin: LIT Verlag.

Kraus, George. 1969. Chinese Laborers and the Construction of the Central Pacific. *Utah Historical Quarterly* 37 (1): 41–57.

Krech, Shepard, III. 1999. *The Ecological Indian: Myth and History.* New York: Norton.

Leopold, Aldo. 1949. *A Sand County Almanac.* Oxford: Oxford University Press.

Lewis, David R. 1995. Native Americans and the Environment: A Survey of Twentieth Century Issues. *American Indian Quarterly* 19: 423–450.

Lim, Shirley. 2006. The Native and the Diasporic: Owning America in Native American and Asian American Literatures. *Women's Studies Quarterly* 34 (1-2): 295–308.

Linton, Patricia. 1994. 'What Stories the Wind Would Tell': Representation and Appropriation in Maxine Hong Kingston's *China Men. MELUS* 19 (4): 37–48.

Ludwig, Sämi, and Nicoleta Alexoae-Zagni, eds. 2013. *On the Legacy of Maxine Hong Kingston: The Mulhouse Book.* Berlin: LIT Verlag.

Lynch, Tom, Cheryll Glotfelty, and Karla Armbruster, eds. 2012. *The Bioregional Imagination.* Athens: University of Georgia Press.

Manes, Christopher. 1996. Nature and Silence. In *The Ecocriticism Reader: Landmarks in Literary Ecology*, ed. Cheryll Glotfelty and Karla Armbruster, 15–29. Athens: University of Georgia Press.

Martin, Daniel J. 2007. Lynching Sites: Where Trauma and Pastoral Collide. In *Coming into Contact: Explorations in Ecocritical Theory and Practice*, eds. Annie Merrill Ingram, Ian Marshall, Daniel J. Philippon, and Adam W. Sweeting, 93–108. Athens: University of Georgia Press.

Marx, Leo. 1964. *The Machine in the Garden: Technology and the Pastoral Ideal in America.* New York: Oxford University Press.

Masumoto, David Mas. 1995. *Epitaph for a Peach: Four Seasons on my Family Farm.* San Francisco: Harper Collins.

McFarland, Sarah E. 2013. Animal Studies, Literary Animals, and Yann Martel's *Life of Pi.* In *The Cambridge Companion to Literature and the Environment*, ed. Louise Westling, 152–166. Cambridge: Cambridge University Press.

Meeker, Joseph. (1974) 1997. *The Comedy of Survival: Literary Ecology and the Play Ethic.* Tucson: University of Arizona Press.

Mihesuah, Devon. 1996. *American Indians: Stereotypes and Reality.* Atlanta: Clarity Press.

Murphy, Patrick. 2009. *Ecocritical Explorations in Literary and Cultural Studies: Fences, Boundaries, and Fields.* Lanham: Lexington Books.

Oatman-Stanford, Hunter. 2018. From Yosemite to Bears Ears: Erasing Native Americans from US National Parks. *Collectors Weekly*, January 26. https://www.collectorsweekly.com/articles/erasing-native-americans-from-national-parks

Outka, Paul. 2008. *Race and Nature from Transcendentalism to the Harlem Renaissance.* New York: Palgrave Macmillan.

Partridge, Jeffrey F.L. 2004. *Aiiieeeee!* and the Asian American Literary Movement: A Conversation with Shawn Wong. *MELUS* 29 (3-4): 91–102.

Pfaff, Timothy. 1980. Talk With Mrs. Kingston. *New York Times Book Review* June 15: 1, 25–27.

Phillips, Dana. 2003. *The Truth of Ecology: Nature, Culture, and Literature in America.* New York: Oxford University Press.

Plumwood, Val. 1993. *Feminism and the Mastery of Nature.* London: Routledge.

Posmentier, Sonya. 2012. The Provision Ground in New York: Claude McKay and the Form of Memory. *American Literature* 84 (2): 273–300. https://doi.org/10.1215/00029831-1587350.

Rigby, Kate. 2015. Ecocriticism. In *Introducing Criticism at the Twenty-First Century*, ed. Julian Wolfreys, 2nd ed., 122–154. Edinburgh: Edinburgh University Press.

Rome, Adam. 2003. 'Give Earth a Chance': The Environmental Movement and the Sixties. *Journal of American History* 90 (2): 525–554.

Ruffin, Kimberly N. 2010. *Black on Earth: African American Ecoliterary Traditions.* Athens: University of Georgia Press.

Sakurai, Gail. 1993/1995. The Politics of Possession: Negotiating Identities in American in Disguise, Homebase, and Farewell to Manzanar. In *Privileging Positions: The Sites of Asian American Studies*, ed. Gary Y. Okihiro, 157–170. Pullman: Washington University Press.

Schweninger, Lee. 2007. To Name Is To Claim, or Remembering Place: Native American Writers Reclaim the Northeast. In *Coming into Contact: Explorations in Ecocritical Theory and Practice*, eds. Annie Merrill Ingram, Ian Marshall, Daniel J. Philippon, and Adam W. Sweeting, 76–92. Athens: University of Georgia Press.

———. 2008. *Listening to the Land: Native American Literary Responses to the Landscape. Athens.* University of Georgia Press.

Silko, Leslie Marmon. 1986. *Ceremony.* New York: Penguin.

———. (1986) 1996. Landscape, History, and Pueblo Imagination. In *The Ecocriticism Reader: Landmarks in Literary Ecology*, eds. Cheryll Glotfelty and Karla Armbruster, 264–75. Athens: University of Georgia Press.

Simal, Begoña. 2013. Claiming America through Land Empathy: An Ecocritical Approach to Maxine Hong Kingston's *China Men.* In *On the Legacy of Maxine Hong Kingston: The Mulhouse Book*, ed. Sämi Ludwig and Nicoleta Alexoae-Zagni, 95–111. Berlin: LIT Verlag.

Skeldon, Ronald. 1996. Migration from China. *Journal of International Affairs* 49 (2): 434–455.

Sledge, Linda Ching. 1980. Maxine Kingston's *China Men*: The Family Historian as Epic Poet. *MELUS* 7 (4): 3–22.

Smith, Kimberly K. 2007. *African American Environmental Thought: Foundations*. Lawrence: University Press of Kansas.

Smith, Theresa S., and Jill M. Fiore. 2010. Landscape as Narrative, Narrative as Landscape. *Studies in American Indian Literatures* 22 (4): 58–80.

Sohn, Stephen H., Paul Lai, and Donald C. Goellnicht. 2010. Introduction: Theorizing Asian American Fiction. *Modern Fiction Studies* 56 (1): 1–18.

Takaki, Ronald. (1989) 1998. *Strangers from a Different Shore: A History of Asian Americans*. Boston: Little, Brown.

Wald, Alan. 1987. Theorizing Cultural Difference: A Critique of the 'Ethnicity School'. *MELUS* 14 (2): 21–33.

Wald, Sarah D. 2016. *The Nature of California: Race, Citizenship, and Farming since the Dust Bowl*. Seattle: University of Washington Press.

Warrior, Robert Allen, and Paul Chaat Smith. 1996. *Like a Hurricane: The Indian Movement from Alcatraz to Wounded Knee*. New York: The New Press.

Weaver, Jace. 1997. *That the People Might Live: Native American Literatures and Native American Community*. Oxford: Oxford University Press.

Weil, Kari. 2010. A Report on the Animal Turn. *Differences* 21 (2): 1–23. https://doi.org/10.1215/10407391-2010-001.

———. 2012. *Thinking Animals: Why Animal Studies Now?* New York: Columbia University Press.

White, Richard. 1995. Are you an Environmentalist or Do You Work for a Living?: Work and Nature. In *Uncommon Ground: Rethinking the Human Place in Nature*, ed. William Cronon, 171–185. New York: Norton.

Williams, Raymond. 1973. *The Country and the City*. Oxford: Oxford University Press.

———. 1980. Ideas of Nature. In *Culture and Materialism: Selected Essays*, 67–85. London: Verso.

Wolfe, Patrick. 2006. Settler Colonialism and the Elimination of the Native. *Journal of Genocide Research* 8 (4): 387–409.

Wong, Shawn. (1979) 1991. *Homebase*. New York: Plume.

Wong, Sau-Ling Cynthia. 1993. *Reading Asian American Literature: From Necessity to Extravagance*. Princeton: Princeton University Press.

———. 1995. Denationalization Reconsidered: Asian American Cultural Criticism at a Theoretical Crossroads. *Amerasia Journal* 21 (1–2): 1–27.

Yamashita, Karen Tei. 1990. *Through the Arc of the Rain Forest*. Minneapolis: Coffee House Press.

Yang, Caroline H. 2010. Indispensable Labor: The Worker as a Category of Critique in *China Men*. *MFS: Modern Fiction Studies* 56 (1): 63–89.

CHAPTER 5

Cultivating the Anti-campo: An Environmental Reading of "Internment Literature"

After the attack on Pearl Harbor, in December 1941, American authorities confined a particular racialized minority to a limited, barbed-fenced space. The ostracism and seclusion of an entire community, purportedly done in order to guarantee the defense of US territory, uprooted thousands of people of Japanese ancestry from the West Coast.[1] The result was not only social upheaval, most noticeable in the shaken internal structure of those families, but also a temporary or, in many cases, permanent territorial displacement. From 1942 to 1945, the constrained territory where the 110,000 Japanese American "evacuees" were forced to live was, to employ a double oxymoron, excessively insufficient and invariably different from the one they were used to.[2] Nevertheless, it is my contention that many internees managed to build a special bond with that imposed environment, becoming actively entangled with it.

Until the beginning of the twenty-first century, as Anna Hosticka Tamura notes, most studies of the Japanese American internment experience had depicted "the camp landscapes as cruel places comprised of tar-paper barracks, surrounded by barbed wire fences and watchtowers, and located in remote and desolate areas" (2004, 1; see 19–20). While these facts were true of most camps, they failed to conjure up a complete picture on the internment. Most relevant for the purposes of this chapter, such a description left out the crucial element that made life bearable for many of the internees: the adaptation to and transformation of the natural environment they had been forced to inhabit. In her *Disorientation and*

© The Author(s) 2020
B. Simal-González, *Ecocriticism and Asian American Literature*,
Literatures, Cultures, and the Environment,
https://doi.org/10.1007/978-3-030-35618-7_5

Reorientation: The American Landscape Discovered from the West, published in 1992, Patricia Nelson Limerick anticipated the current trend that underscores the agency of Japanese American prisoners in the camps, particularly the interaction with their new environment. Instead of seeing the internees' gardening and landscaping activities as "a symbol of compliance," Limerick claimed that those internment camp gardens loomed as "a symbol of defiance, a visible statement of unbroken will" (1992, 1046). Building on Limerick's motif of "defiant gardens," Kenneth Helphand (1997, 2006) contributed to shifting the focus away from the facts and signs of confinement, like fences or watchtowers, and toward the feats of landscaping and gardening that the Japanese American internees managed to accomplish in just a few years. Following in the steps of Limerick's and Helphand's pioneering efforts, recent scholarship by environmental historians has tried to draw a more comprehensive and accurate picture of life in the Japanese American concentration camps, often using photographs, testimonies, diaries, and life narratives.[3] More often than not, these scholars resort to Japanese American autobiographical texts as sources providing invaluable historical evidence of the internment experience while also underscoring the symbolic potential of the internees' interaction with the new environment, especially through landscaping and gardening.[4]

While the historical and sociopolitical reading of "defiant gardens" outlined above continues to be relevant, the present study attempts to complement such an interpretation with an analysis of internment narratives *qua* literary works. Within internment literature, it is worth comparing those books, like Monica Sone's *Nisei Daughter* or Miné Okubo's *Citizen 13660*, which were composed during or soon after the experience of internment,[5] with retrospective accounts, probably more reflective in nature. This second group of narratives, including Jeanne Wakatsuki Houston's *Farewell to Manzanar*, Lonny Kaneko's "The Shoyu Kid," or Yoshiko Uchida's *Desert Exile*, was published in the 1970s and early 1980s, when the campaigns for recognition and redress were launched, in a favorable atmosphere that coincided with the Asian American movement. To these two types of internment literature, we would have to add those narratives published in the twenty-first century, including novels like Julie Otsuka's *When the Emperor Was Divine* (2002), Perry Miyake's *21st Century Manzanar* (2002), or John Hamamura's *Color of the Sea* (2006), as well as books for children and young adults, like Amy Lee-Tai's *A Place Where Sunflowers Grow* (2006) and Cynthia Kadohata's *Weedflower* (2006, 2008 Kindle edition). Among these books, I have chosen to focus on four representative narratives from different periods, which allows me to engage in a transhistorical analysis: Sone's *Nisei Daughter*, which came out in 1953, Houston's *Farewell to Manzanar* from 1973, and Miyake's *21st Century Manzanar* and Kadohata's *Weedflower*, both published in the first decade of the new millennium.[6]

In "Fatal Contiguities" (2011), Hsuan L. Hsu wonders about the role of literary analysis in teasing out the implications of spatial segregation in the United States and of environmental racism in general. At one point, the critic compares "aesthetic" and literary analysis with the sociological orientation more commonly found in studies of environmental racism, and he concludes that while environmental justice studies focus on sociological data in order to confirm the existence of environmental racism, literary criticism is unique in that it "productively foregrounds the conditions for perceiving (or not perceiving) the mechanisms and consequences of" such racist attitudes (Hsu 2011, 147). By examining the rhetorical strategies devised by these four authors, therefore, I intend to add a new dimension, literary and ecocritical, which is often missing in historical and sociological studies.[7]

Artistic reimaginings of the environment not only proffer the expected aesthetic pleasure but also serve another purpose: that of hinting at previously ignored connections. This is even more evident in the case of neo-internment fiction like Miyake's *21st Century Manzanar*, which projects the internment experience into an imagined future. Until quite recently, as I have argued elsewhere (Simal 2009), the power of nonmimetic literature has been underestimated in environmental criticism. And this is despite the fact that post-mimetic narrative modes like speculative fiction and magical realism constitute privileged instruments for exploring the multifaceted nature of "reality." The metaphorical latitude of nonmimetic modes of fiction allows not only for a deeper reflection on things past but also for a better projection of possible futures.

As literary artifacts, internment narratives resort to both metaphorical and metonymic tropes. While metaphorical strategies can become a valuable tool for imaginative empowerment and insightful analysis, the trope of metonymy can likewise prove intriguingly fruitful. What is most relevant for our purposes is the "embeddedness" of such rhetorical devices in the very environment in which the narratives are set and which they try to (re)construct. Hsu cogently claims that the trope of "metonymy may play a productive role in aesthetic attempts to represent local instances of environmental injustice": because of its "local and material" nature, it "makes visible the fatal contiguities—the effects of environmental risk factors on bodies, minds, social relations, and lived space—experienced in places that have been abandoned or deliberately blighted by deindustrialization, urban 'redevelopment,' and other reconfigurations of social space" (164). Therefore, the exploration of the ways in which both metonymy and metaphor are deployed in these internment narratives will be crucial for our ecocritical analysis.

Even though I will try to avoid the excessive attention to referentiality that plague most studies of internment literature, I will still need to grapple with the historical events that prompted these narratives: the massive incarceration of Japanese Americans during WW2. My primary aim, however, will be to tease out the ways in which these texts attempt to make sense of those events, a project for which a biopolitical-ecocritical lens will prove invaluable.

1   From Camp to Campo[8]

By signing the Executive Order 9066, on February 19, 1942, Franklin Delano Roosevelt paved the way for the removal and imprisonment of more than 110,000 Japanese Americans, two-thirds of whom had already been born in the United States and were therefore American citizens. There is some debate concerning the most accurate name for the camps to which Japanese Americans were sent. Officially (and euphemistically) called "relocation centers" and managed by the War Relocation Authority (WRA), over time such centers came to be referred to as "internment camps" for a variety of reasons.[9] By the 1970s, that was the phrase most often favored by the government, historians, and even writers who had been former internees like Sone and Okubo. David Takeda, for instance, the Sansei protagonist of Miyake's novel, describes how his parents and grandparents, when reminiscing about their incarceration experience during WW2, omitted the type of "camp" they had been sent to:

> His folks and other Niseis just called it camp.
> "We met in camp," they'd say. "We were in camp together."
> "Our families lived next to each other in camp," they'd say.
> Kids all thought they were talking about summer camp until they got to college and found out their parent's camp fit Webster's definition of a concentration camp. (Miyake 2002, 1–2)

As the quotation above confirms, ambiguity and euphemism plagued the first attempts to name the historical phenomenon: Japanese Americans were not "campers" but prisoners, inmates of concentration camps.[10] The use of the phrase "concentration camp" immediately evokes a certain parallelism between the Japanese American experience and that of the prisoners in the Nazi camps. In fact, it is only with the benefit of hindsight that we realize that what went on in the American "internment" camps was qualitatively different from what happened in Nazi extermination camps. It is important to remember that, just as their counterparts in European concentration camps, the Japanese American internees had no clear idea of how long and for what real purposes they had been placed in those prison-like enclosures. In fact, one of the reasons adduced to keep the entire racialized community in such camps was to turn them into "hostages to make sure Japan would not act in a hostile manner toward the United States," as Kadohata explains in *Weedflower* (17). This implied that should Japan attack any US interests, these hostages could be sacrificed.

This sense of uncertainty appears both in novels and in autobiographical accounts like *Farewell to Manzanar*, penned decades after the events by James and Jeanne W. Houston, herself a former internee. Even Jeanne, whose memories are somehow filtered and buffered by the knowledge of what had actually happened, recognizes the "anguished" months that many Japanese American families spent trying to reunite (Houston and Houston 1973, 18). After Jeanne's father is arrested and taken away, soon after the attack, her mother and the younger children have to move to a new neighborhood, but they are soon forced to leave. For months there had been contradictory rumors and Japanese American families were at a loss as to what awaited them, although more and more news arrived of a collective evacuation: "There was a lot of talk about internment, or moving inland, or something like that for all Japanese Americans. I remember my brothers sitting around the table talking very intently about what we were going to do, how we would keep the family together" (15, cf. Sone 1953, 160–161). Even after incarceration was a fact, the duration of their internment and what life in the camps would turn out to be like still remained a mystery.

It is in fiction, however, that this feeling of uncertainty is most poignantly conveyed and most effectively conjured up. In Miyake's neo-internment novel, a pervasive lack of control and a constant sense of danger saturate not only the dangerous trip out of LA, where one of the Takeda brothers dies, but also the ironically named "sweet home" in

Manzanar. Miyake often employs suspense techniques to keep the reader's attention while skillfully conveying the uncertainty that the evacuees felt at the time. Both suspense and empathy with the anguished prisoners increase after Chapter 26, which ends with a last enigmatic word, the Plan. The supervisor of the Manzanar concentration camp, Lillian, has just discovered that a teenager has got pregnant in camp and, because of that, she feels disappointed in her "beloved" Japanese guests, from whom all she expected "was to act more Japanese" (Miyake 2002, 160). Frustrated, Lillian starts to devise "a way to nip this problem in the bud, so to speak," and concludes with a brief reflection that the author highlights by placing it at the end of the chapter and in a separate paragraph: "If worse came to worse, there was always the Plan" (161). The last word sounds all the alarms. Both the use of an initial capital letter and the choice of the generic noun are highly reminiscent of the Final Solution, the infamous Nazi plan of extermination.[11]

In a very different novel, *Weedflower*, which recounts twelve-year-old Sumiko's experiences during the war as seen through her eyes, Kadohata captures the precise moment when the girl realizes not only that nothing will ever be the same but also that her family has lost control over their lives once and for all. Sumiko, as many other West Coast Japanese Americans, has had to leave everything behind due to the evacuation orders. As soon as she and her family arrive at the assembly center, a horse racetrack, a man tells them they have been assigned "to Section Seven, Stable Four," their new temporary home; what surprises and at the same time irks sensitive Sumiko is that the man announces their new abode "as if everything were perfectly normal" (Kadohata 2006, 85). Upon arriving at their new "home," Sumiko's five-year-old brother, Tak-Tak, takes her hand and asks her if they will share the stable with horses: "'No,' she said, although she honestly didn't know. In fact, she thought *anything was possible now*" (85, emphasis added). These and other examples confirm that uncertainty plagued the Japanese American evacuees as much as it plagued the inmates of the concentration camps in Europe at about the same historical time.

The racist nature of American concentration camps is also undeniable, especially when we compare the situation of the Japanese with that of other "enemy aliens" who were not racialized, Germans and Italians, and who consequently did not suffer collective incarceration as an ethnic group.[12] If anti-Japanese sentiment had timidly emerged at the turn of the century, in the 1920s and 1930s the hatred became conspicuous (Wald

2016, 77–78), as Japanese immigrants grew more numerous and prominent in the Westernmost states, often as farmers in fertile lands.[13] In the atmosphere of popular paranoia that followed the attack on Pearl Harbor, Americans of Japanese ancestry became what Giorgio Agamben calls the *homo sacer*, the individuals that could be dispensed with for the sake of the nation, regardless of their loyalty and citizenship. The WRA could do what they wished with these dispensable bodies, who were forcibly uprooted, tagged as objects (Miyake 2002, 92; Houston and Houston 1973, 17), and sent to remote concentration camps. Once more, it becomes evident that the discourse that sustained the very existence of concentration camps, as theorized by Agamben, was common to both Japanese American and Nazi camps, even if their racial projects were different (Lee, para. 44). In the Nazi concentration camps, the political understanding of human beings as citizens had been annulled and people had been reduced to what Agamben describes as *nuda vita* or bare life, which could therefore be suppressed without committing homicide (1995, 154).[14] Much the same happened in the American concentration camps: here, too, citizenship rights had been denied by the state authorities and internees had been reduced to bare life. Once Japanese Americans were forcibly "relocated" to stables and barracks, as Sumiko says in *Weedflower*, "anything was possible."

Among contemporary philosophers, it is probably Agamben who has developed the most influential discourse regarding concentration camps. In his *Homo Sacer: il potere sovrano e la nuda vita* (1995), Agamben reiterates Walter Benjamin's conviction that, in our times, the state of exception has become widespread, that is, not an *exception* any more (63). For the Italian philosopher the concentration camp is precisely the "space" that opens when the state of exception starts to become the norm (188).[15] And that biopolitical regime that enabled the concentration camps to exist during WW2 continues to underlie the present. In other words, the camp does not constitute an aberrant anomaly of the past but epitomizes the very political structures of modernity: it constitutes "il paradigma nascosto dello spazio politico della modernità" (135), the new biopolitical *nomos* of the entire planet (185, 198). According to Agamben's theory, then, we have all become virtual *homines sacri* (123, 127), human beings who can be eliminated with total impunity even if not ritually immolated (91).

What is even more relevant for our purposes is that the concentration camp, as stated in Agamben's thesis, does not correspond exclusively to the Nazi death camps we all have in mind, but to any space where bare life and norm become indistinguishable from each other.[16] Among the examples of such camps that the philosopher cites is the football stadium in Bari where illegal Albanian immigrants were temporarily interned before repatriation in 1991, or the so-called *zones d'attentes* or waiting areas in French airports, designed for those foreigners applying for refugee status (Agamben 1995, 195). To Agamben's examples, we can add the assembly centers and "relocation camps" for Japanese Americans during WW2 or, more recently, the Guantánamo prison erected after the 9/11 attacks. In all of these places, from the obviously restrictive to the apparently "innocuous" ones, we witness a paradoxical reversal: "A un ordinamento senza localizzazione (lo stato di eccezione, in cui la legge è sospesa) corrisponde ora una localizzazione senza ordinamento (il campo, come spazio permanente di eccezione)" (197).

Therefore, the state of exception, the deprivation of political rights, if not of life itself, is common to all concentration camps, and the Japanese American internment experience proves paradigmatic in this respect.[17] Deprived of freedom, Japanese Americans, many of them US citizens *de jure*, ceased to be so de facto. Purportedly "humane" treatment did not detract from the process of denaturalization and dehumanization that the internees underwent. After Pearl Harbor, Japanese Americans living in the Western states were removed from their homes, temporarily placed in former horse stables, and eventually sent to pen-like barracks, thus reinforcing the animality to which they had been reduced, the bare life they had come to signify. Furthermore, Japanese American concentration camps did not occur in a vacuum, but involved the materiality of environment, which could be read either as reinforcing the "punishment" or as providing literal and figurative escape routes. The environment, I would argue, was instrumental in rendering possible a partial metamorphosis because, in Japanese American concentration camps, the status of the inmates oscillated between the impotent *homo sacer* and the more powerful figures of the *homo faber*, *homo agricola*, and *homo ecologicus*.

2   FROM *HOMO SACER* TO *HOMO ECOLOGICUS*

By their forced internment in concentration camps, as seen above, Japanese Americans were reduced to the role of Agamben's *homo sacer*. Nonetheless, despite the general resignation with which the evacuees seemed to accept their lot, Japanese Americans nonetheless managed to use the remnants of their limited agency to undermine their status as the expendable *homo sacer*. It was mainly through various artistic forms including crafts (*homo faber*) and through their conscious interaction with their new natural environment (*homo agricola*) that Japanese Americans tried to transcend their new position as discarded beings that had only the right to "bare life." In what follows I will show how these second-class citizens claimed agency by becoming artists and gardeners and, ultimately, emerging as the figure of the *homo ecologicus*.

Japanese American internment narratives offer abundant examples of both agricultural and artistic endeavors.[18] Some of these narratives are in fact the direct result of literary and graphic art developed during the years of the internment, most notably Okubo's *Citizen 13660*. Others, although written many years after the incarceration, bear witness to the creative impulse of the internees, who often resorted to carving, drawing, painting, or writing haikus, among other artistic activities. Such activities served Japanese Americans as a powerful reminder that they were more than *homines sacri*, that they could aspire to become a veritable *homo faber*, capable of producing, of creating something out of the very limited resources they encountered in the camps. More relevant to our purposes, the internees soon developed an urge to interact with the new environment in which they had been forced to live. If this desire was already noticeable in the assembly centers to which they were initially sent, it became much more conspicuous in the camps themselves, which they struggled to turn into their home for (at least) the duration of the war. For that purpose, Japanese Americans engaged in gardening and landscaping in virtually all the internment camps they were sent to. Some even managed to temporarily trespass the boundaries of the camp in order to work the adjacent fields or were allowed to travel to more distant areas where agricultural help was needed (Hayashi 2007, 87–93). In this way, the Japanese Americans who became gardeners and landscapers of their own accord managed to transcend their reduced status of *homo sacer* and gradually metamorphosed into what I have called *homo agricola*.[19] The process whereby Japanese Americans became *homines agricolae* was by no means smooth or easy. In most cases it involved a sudden uprooting that compelled many families to leave behind their familiar environment and many of their loved ones. In their portrayal of the new spaces where the characters were forced to live, internment narratives highlight two types of interaction with the new environment: one that involves animalization and another that pertains to what I have called environment shock.

2.1   Animal Spaces and Environment Shock

Before being relocated to the permanent concentration camps, most Japanese Americans were taken to temporary assembly centers that were still in the West Coast area, relatively close to home.[20] In depicting these temporary dwellings, both Sone and Kadohata denounce the animalization that Japanese Americans suffered at the hands of their government in a direct or oblique way. In *Nisei Daughter*, when the bus arrives at their assembly center, young Monica openly describes the shock at being taken to a chicken farm and being installed in former pens (Sone 1953, 173; cf. Miyake 2002, 162). On some occasions, though, Sone prefers to use sarcasm, as when the narrator ironizes about the fact that when the evacuation orders came, there seemed to be more concern for animals than for human beings. According to the newspapers, one of the assembly centers in Washington State was to be the fairgrounds at Puyallup: "The article apologetically assured the public that the camp would be temporary and that the Japanese would be removed from the fairgrounds and parking lots in time for the opening of the annual State Fair. It neglected to say where we might be at the time when those fine breeds of Holstein cattle and Yorkshire hogs would be proudly wearing their blue satin ribbons" (Sone 1953, 160–161). If the complaint of reverse speciesism were not clear in the narrator's ironic comments, a later scene confirms this accusation. As she is awaiting evacuation, Monica looks at herself in the mirror and has trouble recognizing her figure, which has undergone a peculiar theriomorphic transformation: with the "white pasteboard tag" and dressed in many layers of clothes, she "looked like a cross between a Japanese and a fuzzy bear" (167). In addition to the animal presence in her altered physical appearance, her outer layer, animality seeps into a more profound layer, obstructing or eliminating her civil rights as a human being. While packing for their yet unknown destination, Monica realizes that, from that moment on, she has to be prepared for everything: "I was certain this was going to be a case of sheer *animal* survival" (161; emphasis added), of being reduced to bare life. Examples such as these show that Japanese Americans who, like Sone herself, went through this experience were quick to perceive that, by having been transferred to stables, they had come to occupy both real and symbolic animal spaces. Replacing nonhuman animals automatically rendered these people inferior in the social hierarchy. The authorities' choice of temporary centers of assembly for Japanese Americans, therefore, was highly significant. In Takashi Fujitani's apt

words, it bore witness to "the easy slippage from metaphor to state practice" (2007, 29–30): Japanese Americans had suddenly undergone the
shift from figurative animalization to the literalization of the demeaning metaphor.

However, metaphor is not the only trope these narratives use when
describing and denouncing animalization. A common complaint among
the evacuees in both fiction and autobiography was the persistent nonhuman smell: since most of these improvised quarters "had been designed to
house animals" and not human beings, "the stench of animals remained"
(Helphand 2006, 157; cf. Limerick 1992, 1037). The narrator in
Kadohata's *Weedflower*, in depicting the stable, highlights both the smell
and the stunned reaction that every member of the family has on seeing
their new "apartment": "Ichiro stepped gingerly inside. Sumiko followed
and then the others. *Nobody spoke*. It looked like a two-horse stable. It
stank of manure and chlorine" (86; emphasis added). The persistent smell
that lingered in the room despite the disinfectant reminded the new "residents" that this was an "animal space" in more ways than one. In these
narratives, the trace of animality became physically attached to the new
inhabitants. Therefore, not only the metaphorical connotations of the
stable homes but also that material trace of animality seemed to permeate
the new inhabitants. Thus, the trope of metonymy enhances the environmental reading of these narratives: the temporal and spatial contiguity of
animals in the form of their solid and less solid detritus (manure, smell)
further reinforces the process of "therio-metamorphosis" of Japanese
Americans.[21]

Without forgetting the dangers of an uncritical understanding of animalization, as explored in Chap. 3, I venture to claim that animal spaces
demoted the evacuees/internees in the social hierarchy while they also
reminded them of the very real animals they had had to leave behind.
Nonhuman animals constitute a significant presence/absence in internment literature. Just as "animal relatives" had played an important role in
Edith Eaton's fiction, family animals, especially dogs, had also been part of
Japanese American life for decades.[22] The image that best conjures up
what these nonhuman animals meant for the displaced Japanese Americans
is the diorama described in Kadohata's *Weedflower*. When sensitive Sumiko
arrives at the temporary center to which they have just been evacuated, she
is appalled. In its simplified bareness, the assembly center reminds her of
the dioramas they had once created at school: "each group had made a
diorama of life in a different country. ... The teacher said all the kids did a

great job. But now Sumiko thought that if you had enlarged all the diora-mas, there would have been things missing, like curtains, pets, and gar-dens. *Details*. Those details were also missing from the assembly center" (Kadohata 2006, 91). By highlighting the word "details" both graphi-cally—through italics—and syntactically—making it stand on its own—Kadohata underlines the centrality and importance of such "details," most prominent among them nonhuman creatures.[23]

Despite the ban against keeping animal companions in the concentra-tion camps, some internees eventually managed to keep small animals in the barracks or they made new animal friends in the camp. It goes without saying that the most symbolically resonant companion that can be kept in a prison is a bird, the emblem of freedom—think of classic prison films like *Birdman of Alcatraz* or *The Shawshank Redemption*. As befits a prison nar-rative, Sone's *Nisei Daughter* also features an inmate in love with birds. One day Monica visits the old man in his untidy barracks room. On seeing Monica intrigued by the "odd" covered cage she finds there, Mr. Sawada explains, with a broad smile, that "this is where Shozo lives" and, uncover-ing the cage, he shows Monica "a disgruntled back crow" (Sone 1953, 233–234). Lonely Mr. Sawada had sought out the bird and grown attached to him: "For hours I sat out there on the prairie before I could persuade him to light on my shoulder. Since then we've become good friends. Sometimes I let him fly around in here" (234). Mr. Sawada's crow consti-tutes an apt metaphor for the internees' own situation: his movements, like theirs, are severely restricted. The presence of the crow in a cage in a concentration camp, in a figurative *mise-en-abîme*, points at the multiple and varying degrees of incarceration, while it ultimately functions as a hyperbolic quotation underscoring the insidious nature of imprisoning liv-ing creatures, both human and nonhuman.

This discussion of nonhuman animals, especially those we tend to bond with, those that we perceive as "our pets," inevitably leads us to ponder not only on speciesism, on the belief in the superiority of humans over other animals, but also on the internal hierarchies existing within the non-human animal kingdom.[24] Although many of us establish intimacies with, and deeply care for, animals like dogs or cats, very few humans establish such close relationships with rats, mosquitoes, or worms. As we saw in earlier chapters, speciesism can be explained by the fact that both tradi-tional and contemporary ethical projects tend to invoke the human/non-human divide as inviolable, as "hyperseparated," to use Plumwood's term. Notwithstanding the ethical extension advocated by environmental philosophers like Leopold or Plumwood, George Orwell's dictum is still literally true: "All animals are equal, but some are more equal than others." What makes this internal speciesism so recalcitrant? What makes human "biophilia" (Wilson 1984) so selective?[25]

One can try and find the answer to that question in our scientific, taxonomic tradition, starting with Linnaeus, but one can also explore the premises of contemporary ethics and of contemporary affects. Emmanuel Levinas, one of the most influential ethical philosophers of the twentieth century, bases his theory of ethics on the appeal to "the face of the other," initially construed as the human other. Even if we stretch and broaden Levinas's concept of the interpellating face/other to include great apes or family animals, this type of ethical extension still leaves out "faceless" creatures. In most societies, humans find it relatively easy to relate to other mammals like cats and dogs (we belong, after all, to the same class of vertebrates), but what about invertebrates? What about headless/faceless animals like starfish or corals? Are we as classist in nature as in society? Some would argue that human attitudes to other living beings, especially to other animal species, respond to a conscious, rational hierarchization of species according to whether these nonhuman animals have more or less capacity to feel pain or whether they are more or less complex organisms. However, our differing reactions to different animal species may ultimately respond less to scientific motives than to affective ones: it is physical resemblance and familiarity that, more often than not, determines who becomes our animal relative and who becomes family.

In their initial description of the material context of "relocation," internment narratives not only highlight the animality to which Japanese Americans are reduced in their new environment, but they also depict an equally disorienting experience, which can be described as landscape shock, or, more broadly, environment shock.[26] While this shock was more conspicuous among those Japanese Americans who were evacuated directly to camp, with no temporary sojourn in an assembly center, it is also noticeable in those who did go through the assembly centers.[27] The very manner in which the journey to the concentration camp was orchestrated by the military authorities both underscored the Japanese Americans' animalized imprisonment and guaranteed their landscape shock. Like cattle, Japanese Americans were kept in literal and metaphorical darkness. Furthermore, this figurative blindfolding reinforced the image of "captives" and alien enemies that Japanese Americans had seen emerge and consolidate in the preceding months (Houston and Houston 1973, 16–17; Kadohata 2006, 34–37;

Sone 1953, 157). Kadohata's *Weedflower* narrates how, when the train finally "pulled into the town of Parker, Arizona," it was extremely hot inside the cars/wagons, "like when you light *an oven* and it keeps getting hotter. One man said cheerfully, 'It beats being shot!'" (Kadohata 2006, 99; emphasis added). Once, during the night, Sumiko manages to peek out and sees unrecognizable trees: giant cactuses (99). The landscape shock is such that the girl feels she is in another planet. Likewise, in Sone's *Nisei Daughter*, the narrator remembers how, after drinking in "the extravagant beauty … [of] the Hood River region along the Columbia River," dinnertime came and the MP told them to pull down the shades (190–191; cf. Houston 18). The following morning, when she and the other passengers were finally allowed to look through the windows, "the scenery had changed drastically […]. The staring hot sun had parched the earth's *skin* into gray-brown *wrinkles* out of which jagged boulders erupted like *warts*. Wisps of moldy-looking, gray-green sagebrush dotted the land. In the midst of this bare prairie, our train clanked to a standstill" (191; emphasis added). As the highlighted terms show, the personification of the landscape is evident. Filtered as it is through the girl's imagination, the resulting image approximates that of the ugly witch in fairy tales, a figure far from appealing and welcoming, a figure that often conjures up deceit and danger.

If it had not happened before, landscape and climate shock hit in as soon as the evacuees—especially those who had been forced to remain "blind" to their changing environment—arrived at the camp itself. The reason was obvious: eight of the ten camps were set in the desert, while the remaining two were built in rural "swampland" in Arkansas. In the case of the four narratives analyzed in this chapter, the final destination of the evacuees would all be in (semi)arid areas, in the camps known as Poston (Kadohata), Minidoka (Sone), and Manzanar (Houston and Miyake). The landscape and climate shock was the logical outcome of such a drastic alteration. For the younger generation of Nisei, who had only known fertile farming lands, some fishing areas along the West Coast, or Western cities like San Francisco, Seattle, or Los Angeles, the bewildering contrast must have been as great as that experienced by the older Issei, the first generation of immigrants or those sent to school in Japan (see Kadohata 2006, 129).[28] In all cases, the new camp environment would be hard to assimilate, whether you had grown in a farm or in a big city: "Desert quiet. Compared to LA, this was outer space" (Miyake 2002, 162).

Environment shock was more than an aesthetic reaction to what the eye could see: it was a multisensorial phenomenon, as we can surmise in Miyake's reference to absolute silence. In addition to sound and sight, other senses were profoundly involved in this first encounter. Drastic temperature changes, for instance, made it difficult for Japanese Americans to adapt to their new arid environment. The extreme climatic conditions, which combined intense heat in summer and unusual cold weather in winter (Sone 1953, 196–197; Miyake 2002, 104), were also interpreted as an aggression on the part of nature. In Houston's memoir, natural elements like the sun are perceived as vampire-like agents. Jeanne remembers how she "would hike past row upon row of black barracks, watching mountains waver through that desert heat, with the sun trying to *dry up my very blood*" (Houston and Houston 1973, 44; emphasis added).

While the rigors of extreme heat or cold beleaguered the camp inmates, it was the sense of touch that was most visibly "assaulted" by sandstorms.[29] It can be said that the abrasive materiality of the new environment literally became a slap in the face for these displaced people, whose first contact with the new "home" was commonly wrapped in dust. In their respective autobiographical accounts, *Farewell to Manzanar* and *Citizen 13660*, Houston and Okubo similarly note the ubiquity of dust. Although these writers describe the inconvenience of having to fight dusty wind, they both use dust as a literary motif of comic relief. Okubo draws the moment of her arrival at Topaz, officially the Central Utah Relocation Project, and describes the welcoming ceremony held amid the dusty wind: "As we stepped out of the bus, we could hear band music and people cheering, but it was impossible to see anything through the dust. ... When we finally battled our way into the safety of the building we looked as if we had fallen into a flour barrel" (Okubo 123; see Uchida 1982, 109). The same joking reference to the flour barrel reappears years later in Houston's autobiographical account:

> We woke early, shivering and coated with dust that had blown up through knotholes. ... I looked at Mama's face ... her eyes scanning everything—bare rafters, walls, dusty kids—... I think the mask of her face would have cracked had not Woody's voice just then come at us through the wall ....
> "Hey!" he yelled. "You guys fall into the same flour barrel as us?"
> "No," Kiyo yelled back. "Ours is full of Japs."
> All of us laughed at this. (Houston and Houston 1973, 23–24)

The way in which the different narrators in internment literature phrase the first encounter with the environment in which the camps are set is particularly illuminating for our ecocritical analysis. In Sone's *Nisei Daughter* there are two such moments: the arrival at the assembly center at Puyallup, ironically called "Camp Harmony," and the first description of the permanent camp at Minidoka, Idaho. The first one emphasizes the degree to which the new inhabitants are engulfed, almost swallowed, by the new environment: "We stumbled out, stunned, dragging our bundles after us. It must have rained hard the night before in Puyallup, for we sank ankle deep intro gray, glutinous mud" (Sone 1953, 173). If this first reference was not enough evidence of the negative connotations accruing to the mud, the narrator goes back to it later when describing everyday life at the assembly center: "We fought a daily battle with the *carnivorous* Puyallup mud. The ground was a vast ocean of mud, and whenever it threatened to dry and cake up the rains came and softened it into a slippery ooze" (180; emphasis added). The metaphorical use of *carnivorous* to qualify the mud, together with the hyperbolic reference to its ocean-like immensity, underlines its danger and inescapability. In contrast, the scene of the arrival to the camp is seemingly described in simpler, less rhetorical terms: "Camp Minidoka was located in … a semiarid region, reclaimed to some extent by an irrigation project with the Minidoka Dam. From where I was sitting, I could see *nothing* but flat prairies, clumps of greasewood, and the jack rabbits. And of course the hundreds and hundreds of barracks" (192; emphasis added). However, even in this apparently straightforward enumeration of what the narrator could first see of the new physical environment, the syntactic choice of a negative structure ("I could see nothing but …") highlights the void in which Japanese Americans would be forced to live. In stark contrast to that nothingness, the endless rows of prison barracks were all the more striking.

In Miyake's *21st Century Manzanar* it is not just David Takeda but also his African American friend, Greg, that focalize the arrival at the arid landscape where the camp is located. What is foregrounded in the description of the new environment—by means of repetition, with variation—is its immense, flat, and barren nature:

> Greg saw it as he came over the hill. *Miles* of *flat* desert and at the bottom, a small road off towards the mountains on the left ….
> Greg turned onto the small road toward the low purple mountains lining the horizon of the *barren flat*land.

... The *dull,* purple mountain rim stretched around them like the edge of *forever, flat, empty, brown,* beige and pale yellow cut in two by the two-lane asphalt highway. (Miyake 2002, 91; emphasis added)

Here, the landscape shock is of such magnitude that David and Greg fail to notice the threatening car approaching them, and the environment thus becomes a figurative accomplice of the attack they suffer immediately afterwards: "if they hadn't been blown away by the immense, barren vastness engulfing them, they might've noticed the brown sedan" (Miyake 2002, 91). Although David and Greg will put the blame on themselves, the last lines also hint at a treacherous landscape that has mesmerized them with its strangeness and sublimity, and whose nondescript color allows the impending danger (the brown sedan) to camouflage itself. Miyake's first laconic portrayal of the Manzanar site, therefore, highlights its sterile flatness and a dull brownness that seems to pervade everything (see Sone 1953, 188).

Like the other three narratives analyzed in this chapter, Kadohata's novel also includes a significant arrival scene where the environment shock is the most prominent feeling among the children, Sumiko and Tak-Tak. This time, however, it is the strange climate rather than the new landscape that overwhelms the characters. When Tak-Tak gets off the train, at Parker, Arizona, it is not the number of soldiers that surprises him, but the heat. His sister feels the same and she can find nothing similar in previous experiences: "Sumiko had never felt anything like it either. Even when she was lighting the fire under the bathtub, the heat was contained in just the area of the fire. ... But here the heat was everywhere, as if there were fire all around them" (Kadohata 2006, 99). Although the corollary remains unsaid, it is implicitly hinted at: the internment experience itself is as inescapable as the heat, and it is not limited to just one area, the camp, but seems to impregnate their whole life as Japanese Americans. Once more, this is reminiscent of Agamben's campo paradigm, which does not just characterize one location but saturates the entire system.

After being escorted into buses, the evacuees are taken to their final destination, the "Colorado River Relocation Center," most commonly known as Poston. After being kept in the dark for the duration of the journey, Sumiko can finally look through the window. To her great disappointment, the child realizes "there was nothing to see, just a few old trees and tangled bush and a lot of light brown dust" (Kadohata 2006, 100). Old enough to have built a mental image of what the desert was going to be like, Sumiko expected to see at least the typical sand dunes. Instead of dunes, however, "this desert was filled with *dry* bushes and *dry* trees growing in *dry dirt*. Tractors moved like huge *dirt*-eating animals" (103). As can be observed in the repetitive cadence, arid dryness and plain dirt replace exotic sand dunes. To the child's eyes, the only redeeming element in the landscape seems to be a range of mountains rising "in the distance," but even these are overshadowed by "a strange, dirty cloud hover[ing] near them" (100). In fact, as the bus gets closer to Poston camp, the strange cloud grows larger and more menacing. Since Sumiko is the main focalizer of the scene, she tries to understand the new phenomenon by resorting to familiar images from her short life: "It looked like brown, boiling water. Closer to the bus she also saw small swirls of dust rise up from the ground like dancing girls. For a moment she thought the temperature had grown so hot that the air was actually boiling. Maybe that was possible" (101). Enveloped in the dust storm, the bus can no longer run and its passengers remain trapped inside for an hour. When the storm finally dies down, they get off the bus and the men help the driver get the vehicle unstuck. What Sumiko sees through her childish eyes conjures up the uncanny nature of the situation: "Some of the men were covered head to toe in dust. They looked as if they'd jumped into a vat of brown paint. ... Nothing seemed real. The world outside was brown. The sky was sort of tan. All the men were brown; even Sumiko in her previously mint green school dress was brown. Everything in the world was brown and very hot" (102). The child's candid perception responds to her need to understand things in simple chromatic-temperature terms. In addition, Sumiko's description of the hot brownness enveloping "everything in the world" foregrounds the extent to which, upon arrival, people start to mimetically blend with the landscape: the men outside the bus and Sumiko herself are now indistinguishable from the natural environment, an environment that has transformed them by wrapping them in homogenizing brownness.

The description of the first contact with the camp in Houston's *Farewell to Manzanar* equally provides useful formal clues for a better understanding of the characters' environment shock. When the bus taking young Jeanne finally arrives at Manzanar, it is suddenly "pelted by what sounded like splattering rain. It wasn't rain. This was my first look at something I would soon know very well, a billowing flurry of dust and sand churned up by the wind through Owens Valley" (Houston and Houston 1973, 18–19; see 72–73). Once more, the way in which the narrator phrases this encounter with the new environment is rather telling: faced with the unknown natural environment, and especially its new climatic conditions, the young girl tries to find a familiar equivalent, rain. The telegraphic sentence "It wasn't rain" not only underscores the unfamiliarity of the new environment, but it indirectly anticipates one of the internees' major concerns and desires: for the desert land to receive some water so that something "green" could grow in that soil. In addition, the hostility implicit in "pelting" and readily attributed to the dust cannot but remind readers of an active agent, as if dust had a mind of its own and was willingly aggressive and ill-disposed toward Japanese Americans, like many American people were at the time. Sone's similar choice of a violent simile to describe the frequent sand storms confirms this interpretation: "We felt as if we were standing in a gigantic sand-mixing machine as the sixty-mile gale lifted the loose earth up into the sky, obliterating everything. Sand filled our mouths and nostrils and stung our faces and hands *like a thousand darting needles*" (Sone 1953, 192, emphasis added; see Miyake 2002, 234). In these narratives, the natural environment comes across as hostile, as actively attacking the newly arrived prisoners. Limerick's fortunate phrase that the WRA and nature had "joined forces to construct landscapes designed to break the spirits of the prisoners" (1041) proves rather accurate in this context: every textual clue seems to confirm the intuition that desert landscapes and dust storms were commonly perceived as a punishment.[30]

Dust, however, not only serves as the emissary of the environment, the one that gives the first "welcoming" slap to the newly arrived residents of the camp; it can also function as the unstable and mobile membrane between the human beings and their new natural surroundings. The fact that the ankle-deep mud that "welcomes" Monica upon arrival threatens to "eat" her alive, as described in *Nisei Daughter* (173, 180), reinforces the oppressive experience of feeling engulfed by the new environment. Due to their insidious intrusiveness, mud and dust also invite metonymic readings, like that captured by Stacy Alaimo's concept of trans-corporeality.[31] As mentioned in Chap. 1, the key premise in Alaimo's eco-theory is that human beings and their environment shape and constitute each other. Since the emergence of our species, but more obviously in the era of the Anthropocene, we have directly altered or indirectly affected the environment in which we live, both locally and globally. At the same time, even though it seems difficult to accept "agency without a subject" (Alaimo 2008, 245), we cannot deny the peculiar agency that material "nature" has. The inextricable co-constitution of human corporeality and its physical surroundings renders ineffectual any attempt to push the natural environment to the background. Trans-corporeality precisely captures the insight that "the human is always intermeshed with the more-than-human world, and underlines the extent to which the substance of the human is ultimately inseparable from 'the environment'" (Alaimo 2010, 2; see Alaimo 2008, 238).

Although Alaimo chooses to illustrate her theory with the negative example of toxic bodies, I contend that a trans-corporeal approach can also show material interpenetration as having a positive effect. Arguably, in the internment narratives under analysis here, dust enables and highlights apparently unwanted trans-corporeal relations. These texts underline the pervasive presence of dust and the impossible task of getting rid of this intrusive environment. I would maintain, however, that the interpenetration with the environment is not always perceived as negatively as in the first encounter with the new landscapes. Soon after arrival, Japanese Americans start a figurative dialogue with the natural environment in which they have been forced to live. In this conversation, the landscape stops being perceived as an enemy and gradually becomes a companion. The trans-corporeal perspective sketched above will help us to grasp how dust and the human body can interact in unexpected ways, as illustrated by Miyake's and Kadohata's narratives.

2.2   A Dialogue with the New Environment

The dynamics of interaction between Japanese American characters and their new environment in internment literature changed over time, allowing for the emergence of different paradigms: domestication, adaptation, or even, in some cases, conservation. As time went on, the initial domesticating impulse was superseded by other forms of "dialogue" with the natural context in which the environment finally "had a say." Although this shift is more fleshed out in internment narratives than in historical records,[32] the new conservationist attitude can also be found in archival documents like the *Minidoka Irrigator* (see Hayashi 2007, 105), where it occasionally transpires that the prisoners' experience had taught them to value nonhuman life, especially plants and trees, more than ever before. Like the old gardener in *Weedflower*, many a Japanese American "stood by his plot of land holding a handful of droopy bean cuttings as if he were holding gold" (Kadohata 2006, 144).

Patterns of Domestication

Despite restrictive legislation—most notably the Alien Land Laws discussed in previous chapters—by the time Pearl Harbor was attacked, in December 1941, Japanese immigrants and their Nisei children had had a long agricultural tradition in America.[33] Although the California Alien Land Law of 1913 had prevented Issei from owning and, more crucially, leasing real estate, including the plots and farms they worked in (Daniels et al. 1986, 171–72; O'Brien and Fujita 1991, 23–26; Wald 2016, 46, 114), the agricultural experience of Japanese Americans was still extensive, first as hired laborers and tenants in other people's farms and later as workers in their own lands, which had become legal property of the American-born Nisei (see Kadohata 2006, 6). In addition, urban Japanese Americans were also familiar with gardening and landscaping, for many of them had been involved in these activities in cities like Los Angeles, San Francisco, or Seattle.[34]

As a consequence, by the 1930s almost two thirds of the Issei and Nisei living on the West Coast worked in "agriculture-related activities" (Helphand 2006, 158).[35] Nevertheless, it is still rather surprising that Japanese Americans started their own gardens as soon as they were "evacuated," even in the temporary assembly centers. There are several reasons behind such a choice. To start with, by cultivating their own vegetables, the Japanese American evacuees increased and improved the food they had to eat in the concentration camp, which had given them severe digestive problems from the outset (Limerick 1992, 1041). Although several narratives mention the poor quality of the food and its negative consequences, like the episodes of general food poisoning and diarrhea mentioned by Sone (178) and Kadohata (88), or the "Manzanar runs" that Houston describes in her autobiography (30–31), it is Miyake's novel that gives it particular significance. Not only does David become the camp's latrine man, even singing a "Benjo Blues" (literally, "toilet blues", Miyake 2002, 95–100; see Houston 31–34), but, more meaningfully, it is the bad food causing them digestive troubles that starts the "rice riot" (Miyake 2002, 219–220), the turning point in the novel.

However, the Japanese American prisoners not only worked the land for the practical purpose of food production, they also engaged in landscaping and ornamental gardening; in other words, they created both "productive space and amenity/contemplative place" (Tamura 2004, 8). Several authors have suggested that, as inheritors of a purportedly nature-oriented culture, Japanese Americans valued both the aesthetic and the practical aspects of their involvement with nature. Still, it remains highly contentious whether the popularity of gardens among the camp prisoners originated in the Japanese cultural legacy, most vividly felt by the first generation.[36] Whatever the case may be, the paradigm in which camp inmates initially worked was one of domestication of nature, similar to the ones we find in Jeffersonian agrarianism (Limerick 1992, Hayashi 2007, Wald 2016), in the notion of middle landscape (Marx 1964, Rowe 1989), or in pioneer discourse (Helphand 2006; Hayashi 2007). Surprisingly, Japanese Americans writing for the camp newspapers like the *Topaz Times* or the *Minidoka Irrigator* specifically referred to the problematic figure of the pioneer.[37] Nevertheless, it must be noted that such newspapers were under the control of government supervisors and authorities who no doubt favored such idealized and optimistic readings.

It may be argued that the environmental discourse shaping the work of Japanese American internees bore a striking resemblance to that of American agrarianism, even though this term has often been misused (Govan 1964) and needs to be redefined under the current ecocritical aegis (Fiskio 2012). Still, Japanese Americans did not imitate so much as rival the American agrarian ideal. As Limerick notes, Asian American gardens, including those "in the concentration camps, as well as the earlier Chinese and Japanese vegetable gardens and farms throughout California, offered a powerful statement that the term 'agrarian dream' could carry the adjectives 'Chinese' and 'Japanese' as well as 'Jeffersonian'" (1992, 1045). Such utopian dreams, however, were to be "tested under very different conditions": when Asian Americans tried to pursue their agrarian ideal, they had to overcome racist prejudices and land laws and, in the case of Japanese Americans, their collective internment in concentration camps during WW2 (Limerick 1992, 1045). It is within this framework that we should understand the earlier references to pioneer discourse on the part of Japanese American prisoners. After all, Japanese Americans were not so much conventional "pioneers," in the Western frontier model of settler colonialism, as sociopolitical trailblazers: it was the "social frontiers" they were called to explore and conquer by overcoming American discrimination.

A relevant question that should be addressed is whether the environmental discourses sketched above transpire in our internment narratives. All the texts analyzed in this chapter, both fiction and autobiographical accounts, mention gardening, landscaping, or farming activities by the Japanese American internees.[38] Even if these internment narratives do not describe specific, technical accounts of agricultural endeavors, they do include signs that the paradigm of traditional agrarianism, with its goal of "taming" nature, was initially the prevalent one in concentration camps.

Although Sone's *Nisei Daughter* describes life before evacuation and in the assembly center in quite some detail, comparatively speaking, it is the book that devotes fewer chapters to the experience of camp life itself. In consequence, it pays relatively less attention to the interaction of prisoners with their natural surroundings than the other internment narratives. Yet, even though the discourse of domestication may not be explicitly endorsed in Sone's memoir, it is implicit in the scene when Monica first hears of Minidoka, the permanent camp where she and her family are going to be sent. In the letter they receive from a Japanese American man who is already in Minidoka, preparing the barracks for human habitation, he describes the "dried-up" landscape where the camp is located, but he is quick to add that "there are compensations. A wonderful wild river roars by like a flood [...] and a gigantic water tank" (Sone 1953, 188–189). What is implied but not said in the letter is that, since water is readily available, the irrigation of these arid lands is possible. After all, it is no coincidence that the camp newspaper be called the *Minidoka Irrigator.*

We also have oblique references to the agrarian ideal in Houston's *Farewell to Manzanar.* At the end of part one of her autobiography, Houston tells of the importance both of cultivating patience and resilience and of valuing nonhuman life. Before leaving for their permanent camp, Jeanne hears an old anthem-like Japanese song: "May thy peaceful reign last long/May it last for thousands of years/Until this tiny stone will grow/Into a massive rock, and the moss/Will cover it deep and thick" (Houston and Houston 1973, 90). While the metaphorical reading of this patriotic song is self-evident in its praise of endurance, the literal meaning of the lines supports the discourse of patient gardening. It is precisely this aspect that Houston fleshes out in the following paragraph:

> The moss is the greenery that, in time, will spring even from a rock. In Japan, before the turn of the century, outside my father's house there stood one of those stone lanterns, with four stubby legs and a small pagoda-life roof. Each morning someone in the household would pour a bucketful of water over his lantern, and after several years a skin of living vegetation began to show on the stone. ... a certain type of mossy lichen with exquisitely tiny white flowers sprinkled in amongst the green. (91)

This Japanese song of endurance is premised on an underlying ethos of domestication. By singing this traditional song, the lesson passed on from one generation to the next is the need to be persistent in one's efforts to shape the environment, much like the philosophy underpinning the bonsai tradition. Yet the attitude toward the environment that transpires in *Farewell to Manzanar* is more ambivalent than that and it does not always conform to domesticating agrarianism.

Environmentally speaking, Manzanar was probably the most paradigmatic and at the same time the most paradoxical of the concentration camps. Set in a desert area near the Sierras, Manzanar allowed for the artistic sublimation found in Ansel Adams's camp photographs. At the same time, Manzanar was more than this apparently desertic landscape. Its name, which means "apple orchard" in Spanish, might have seemed ironic at first. However, it was not without historical reason, because the area where the camp was set had indeed been full of pear and apple trees before the insatiable thirst of the LA urban sprawl had desertified it, as Houston and Houston herself remarks (1973, 95).[39] Still, by February 1943, one year after their imprisonment and soon after the polemical loyalty oath, the Japanese American prisoners were finally filling Manzanar with plants. "Gardens had sprung up everywhere," Houston notes in her memoir, "in the firebreaks, between the rows of barracks—rock gardens, vegetable gardens, cactus and flower gardens" (99). The area near the camp started to produce more than enough vegetables for the Japanese American prisoners. Irrigation had made it possible; and irrigation had been possible itself by the camp internees "stealing back" some of the water that the Californian metropolis was taking from the land. Thus, in attempting to cultivate the Manzanar area, Japanese Americans were both trying to "tame" this natural environment and helping to revert it to what it had previously been. Therefore, in terms of environmental discourse, the prisoners' efforts to fill the camp with greenery and cultivate the adjacent fields could be read as fitting both the domesticating and the conservationist paradigms.

From Failure to Adaptation
Not long after their forced relocation to concentration camps, Japanese American prisoners realized that the new environment was not always amenable to their domesticating efforts. In fact, the productive farm that Houston described in the aforementioned quotations took time, sweat, and, more crucially, a change in attitude. Initially, the Japanese American prisoners had tried to impose their preconceived idea of nature, similar to

"the lost home in the West or the mental picture of the idealized past" (Chen and Yu 2005, 560). In so doing, the internees were just trying to rebuild their lives by reproducing the material context they had left behind.[40] Ultimately, it was that desire to replicate their old environment that moved them to embark upon the impossible task of distilling from this "strange" nature some resemblance to their idealized landscapes. Their initial attempts would soon prove futile; the material conditions, most notably the soil and climate, seemed to conspire against these enterprising prisoners.[41] Unless they learned to adapt to and "cooperate" with their new environment, such a project was doomed to failure.

From our contemporary perspective, we may wonder whether it was reasonable to expect extremely arid landscapes—even the ones that had become desertified due to direct human intervention like Manzanar—to revert to being lush gardens or orchards. As discussed in the preceding analysis of environment shock, most camp internees initially saw the desert as hostile, dangerous, and even bloodsucking.[42] Taking into account the negative connotations that used to be associated with semiarid landscapes, it is no wonder that Japanese American prisoners had not been willing to "cooperate" with their new environment, at least in the early stages. They did not easily accept the new landscapes both because they departed from the ideal and the "normal" and, more importantly, because they had been imposed on them. As my previous analysis of environment shock in internment narratives demonstrates, the choice of the location for the concentration camps was perceived by Japanese Americans as part of the "punishment" doled out by their government.

For some time, then, the camp internees continued to see the desert as an aberrant "anomaly" of nature to be endured or to be tamed and transformed into an orchard. As months went by, however, the prisoners' knowledge of their new environment increased and their perception of these arid landscapes gradually changed. Most internees learned to adapt to their new environment and to cope with this "unfamiliar" nature. If articles in camp newspapers had at first voiced complaints about the "uncooperative landscape," in time the prisoners realized that the qualifying adjective "uncooperative" could apply to them as much as to the landscape. Slowly but surely, the Japanese Americans deracinated from the coastal areas learned to engage in a real dialogue with the interior landscapes of the American West. The shift from a paradigm of mastering and domination to one of cooperation and adaptation became evident in the internees' revision of their landscaping and gardening traditions.

In the end Japanese American prisoners, while not dismissing their cultural legacy—with roots both in Japan and in the American West Coast—learned to adapt it to the new environment.

A good example of such adaptation of gardening traditions can be found in small stone gardens, from the miniature landscapes known as *bonkei*, which the inmates were able to do with the plentiful sand from the desert camps, to larger rock gardens. This small-scale landscaping became one of the most popular activities in camp, perhaps because it was a tradition in which, by using those plants and stones available in their arid surroundings, they learned to adapt to and coexist with the new environment. In *Farewell to Manzanar*, the narrator describes how painting, woodcarving, and especially stone gardening prevented her father from succumbing to alcoholism (Houston and Houston 1973, 97). Rock gardens like this were common not only in the internment camps but also in the special prisons to which some Issei had initially been sent (191). These stone gardens, according to Houston, made imprisonment "more tolerable" (191), so it is no wonder that they would become a fixture of permanent camps, including those set in desert areas. As a result of this increasing interaction with the environment, new types of gardens bloomed in the desert. Many internees, like Jeanne's father, started to use cactuses and desert rocks in order to build their stone gardens; they also used plants fit for arid ecosystems in their vegetable gardens. In "listening to" and trying to cooperate with the new context, Japanese American prisoners were essentially reshaping traditional forms of gardening.

Significantly, in *Farewell to Manzanar*, Houston not only depicts the assuaging effects that such gardens have on depressed inmates like his father, she also believes that the stone gardens built by the internees will be the only remaining physical evidence that the concentration camps ever existed. Houston describes how, as an adult, she goes back to the site where Manzanar camp had once been:

> The old road was disintegrating, split, weedsprung. ... We found concrete slabs where the latrines and shower rooms stood, and irrigation ditches, and, here and there, the small rock arrangements that had once decorated many of the entranceways. ...These rock gardens had *outlived* the barracks and the towers and would surely *outlive* the asphalt road and rusted pipes and shattered slabs of concrete. Each stone was a *mouth*, speaking for a family, for some man who had beautified his doorstep. (190–191; emphasis added)

The fact that Houston sees the stone gardens as the only permanent sign of remembrance endows them with more than symbolic power: the prisoners' creative and respectful interaction with the natural environment becomes central to their experience and to the recovery of their dignity as human beings. The natural elements arranged by people in the shape of rock gardens become active, "speaking" agents, as the anthropomorphic reference to mouths makes clear. Ornamental gardens, landscaping projects, and *bonkei* successfully blended the artistic (*homo faber*) and agricultural (*homo agricola*), and, together, they helped Japanese Americans to overcome their imposed dehumanization as *homo sacer*. What is just as important is that, in such activities, nature figured not merely as a passive instrument for the internees' aesthetic or practical purposes, but as an agent in a reciprocal dialogue, as *natura loquens* (see Chap. 4).

From Knowledge to Conservation

Another way in which the Japanese American internees engaged in a productive dialogue with the environment was through the recreational exploration of their natural surroundings. If gardening and landscaping had allowed the prisoners to reconcile themselves with an apparently hostile environment, their (usually supervised) excursions outside the confines of the camps further ameliorated their relationship with their physical surroundings.[43] The very term recreation combines those two aspects: on the one hand, it conjures up the common understanding of enjoyment and pleasure, and, on the other hand, it evokes the possibility afforded to the internees to create anew, to re-create themselves and their relationship with the natural environment. By exploring their vicinity, Japanese Americans not only came to know their environment better, they also learned to appreciate it. Equally important, these hikes, especially those involving unescorted contacts with nature, gave the prisoners a glimpse of freedom after many months of absolute confinement.

In *Farewell to Manzanar*, the narrator reports that day trips "outside the wire" were finally allowed in 1943 (Houston and Houston 1973, 97), and she makes explicit the aforementioned connection between a sense of liberty, albeit limited, and the first-hand encounter with "wild" nature:

> On weekends we often took hikes beyond the fence. A series of picnic groups and camping sites had been built by internees—clearings, with tables, benches, and toilets. The first was half a mile out, the farthest several miles into the Sierras. As restrictions gradually loosened, *you could measure your*

liberty by how far they'd let you go—to Camp Three with a Caucasian, to Camp Three alone, to Camp Four with a Caucasian, to Camp Four alone. As fourth- and fifth-graders we usually hiked out to Camp One, on the edge of Bair's Creek, where we could wade, collect rocks, and sit on the bank. ... I would always ... sit for an hour next to the stream. ... Water there was the clearest I've ever seen, running right down off the snow. (106; emphasis added)

Any sort of day trip beyond the fence, therefore, had an obvious appeasing or liberatory effect, even when it meant just going to a nearby town, as was the case of Monica's "first day of freedom" in Twin Falls (Sone 1953, 203). However, that feeling of (limited) freedom was especially intense when the excursion allowed for both privacy and a contact with nature. In exploring the natural environment these prisoners not only got a glimpse of the freedom of movement they had lost after Pearl Harbor, but also recovered the right to privacy that the camp's close quarters had taken away from them. In a camp where intimacy was hard to come by, the isolation provided by these outings was more than welcome. Here, the negative, agoraphobic connotations associated with the desert(ed) no-place are superseded with the tranquility provided by the personal encounter with nature, much in the pastoral tradition of the *locus amoenus*, a literary topos to which we will return in the sixth chapter.

On the other hand, the fact that the exploration of the new environment was not spontaneous but "policed" necessarily complicates the way we read such outings. The existence of supervised excursions into "wild nature" thus invites two opposite interpretations: these hikes can be construed as a potential tool to dismantle the paradigm of imprisonment; conversely, they can function as simply a temporary escape valve, a mere ruse on the part of the WRA to appease the increasing unrest among inmates, especially during the difficult months that followed the distribution of the loyalty test. The enjoyment of nature would thus be used to stamp out any attempts at resistance among Japanese Americans: in allowing these outings and "encouraging connections to the surrounding landscape," as Chiang argues, the authorities were trying "to limit Japanese American discontent and diminish feelings of boredom and confinement" (2010, 245).

Therefore, the prisoners' increasing contact with, and knowledge of, their natural surroundings could be interpreted as fostering environmentalist attitudes, which would not be common until decades later. With the obvious exception of Native Americans, the Japanese American internees were among the first Americans to build not only a symbiotic relationship with their new environment but also a relationship that can be properly called conservationist. Though this was already noticeable in some of the historical records and articles published in the 1940s, as previously noted, it is even more conspicuous in the books under analysis, all of which endorse an adaptive and, in some cases, conservationist approach to nature. I have chosen to capture the spirit of such an increasingly environmentalist approach in the symbolic figure of the *homo ecologicus*. In what follows, I will tease out the ways in which these internment narratives manage to foreground ecological attitudes in such a difficult material and historical context.

In Houston's *Farewell to Manzanar*, the centrality of the natural environment is undeniable. Some time after their arrival at the concentration camp, Jeanne relates how it was the incipient orchard that made life bearable for her and her family: "In the spring of 1943 we moved to block 28, right up next to one of the old pear orchards" (Houston and Houston 1973, 95).[44] The narrative foregrounds the fact that nature, synecdochically embodied in "those trees," is responsible for transforming their experience in the concentration camp "from the outrageous to the tolerable" (Houston and Houston 1973, 95). Seen in this light, the family's attempts to keep these trees alive should be read not only as a literary trope—that is, natural elements used metaphorically, as objective correlatives, which they also are—but also as showing a genuine appreciation of nature, an appreciation that usually signals the emergence of an environmentalist attitude.

In the other internment narratives under analysis, there are also certain characters that represent an adaptive, appreciative, and, I would add, ecological approach to the natural environment, most notably Monica and her mother, in Sone's *Nisei Daughter*, and Sumiko, in Kadohata's *Weedflower*. As Monica herself acknowledges (Sone 1953, 175), her mother is wise enough to combine the practicality needed for survival with a purely sensorial enjoyment of the natural environment. It is Mother who first values the aesthetic pleasure to be derived from the mountain landscape near Minidoka, and she does so even before they move to the camp itself, just by hearing the reports about the "magnificent view" from the "Area D barrack doors" (183). Significantly enough, as soon as they arrive at their stable apartments in Puyallup, the assembly center, Mother tries to salvage what most people consider weeds but she sees as worthy flowers. Mother, who, as well as a dietician, is a poet particularly sensitive to beauty, soon reacts by preventing everyone from uprooting them: "Don't anyone pick them. I'm going to cultivate them. ... They're the only beautiful things around here. We could have a garden right in here" (174). Although her son mocks her attitude, family gardens will become a fixture of both temporary centers and permanent concentration camps, and they will be instrumental in keeping some prisoners' balance, especially Issei men like the father in Houston's *Farewell to Manzanar*. Mother's attention to these weedflowers can also be construed in symbolic terms. As happens in Kadohata's novel, whose title highlights the identification of the Nisei protagonist, Sumiko, with the unappreciated weedflower, Mother's insistence on saving such flowers acquires a deep metaphorical resonance. The image is even more potent and effective because of the obvious color analogy: the flowers being discarded as weeds are dandelions. Commonly considered wildflowers and present in several continents, most types of dandelion are fully yellow or, at least, display a yellow "heart," including Asian varieties like the *Taraxacum japonicum*. Therefore, Sone's dandelions, "already pushing their way up through the cracks" of their temporary homes (174), would not only be important in themselves but, once more, as the objective correlative of Japanese Americans who, like those flowers, try to survive in a difficult natural and social environment.[45] If this metaphorical reading might sound dubious, the poem that young Monica improvises right after her mother "defends" the dandelions (174) leaves little room for doubt.

Young Monica's attitude to nature differs from and complements her mother's. The girl's reaction on first hearing of the Idaho camp that would become their final destination is particularly illuminating. In a scene reminiscent of Huck Finn's eager anticipation at the end of Twain's novel, when the boy is setting out for the (Indian) Territory, Monica appears to look forward to the adventure involved in "wild"-sounding Idaho (189). Upon arrival, many evacuees, including Monica, seem excited at the prospect of "feel[ing] the prairie under [their] feet" for the first time (191). The disappointments that follow erode this initial feeling of elation. However, Monica manages to look beyond the "desolate sight" of the prison camp (192), beyond their hardships and confinement, and she offers a rapt description of the prairie nights at Minidoka, a complex eulogy of nature that, as we shall later see, combines the pastoral and the sublime: "The night sky of Idaho was beautiful and friendly. A translucent milk-glass moon hung low against its rich blue thickness, and diamond-cut stars were flung across it from horizon to horizon. This was a strange, gaunt country, fierce in the hot white light of day, but soft and gentle *with a beauty all its own* at night" (Sone 1953, 193, emphasis added). Should we compare this with the initial expectations of this "unknown territory"—heavily influenced by the images Monica had seen in the *National Geographic* magazine—those "dried-up waterholes, runty-looking sagebrush and ugly nests of rattlesnakes" (189) are still recognizable in the emphasis on the bare or skeletal landscape. Still, what stands out in the description of the night scene is the fascination that the new landscape holds for the narrator. Monica has moved from an initial mixture of curiosity and fear toward a feeling of wonder and awe before the ineffable beauty of the Idaho nights.

While Monica and her mother illustrate the environmentalist attitude in Sone's autobiography, in Kadohata's *Weedflower* it is young Sumiko that best embodies the conservationist turn. Narrated as it is from the point of view of young Sumiko, Kadohata's novel subtly contrasts the adults' attitude with the child's particular vision. Growing in her aunt and uncle's flower farm, Sumiko soon learns about different types of flowers, from carnations to "weedflowers": "Local flower farmers called flowers grown in the field *kusabana*—'weedflowers.' Stock were weedflowers that emanated an amazing clovelike fragrance. Of all the flowers her family had ever grown, Sumiko loved them most" (Kadohata 2006, 5). While her uncle was obsessed with achieving the "perfect" carnation and consequently ignored weedflowers like stocks, these were Sumiko's favorite plants. Soon after Uncle's arrest and fearing for her family's future, Sumiko starts spending hours among the weedflowers, the only solace she can find in a world that seems to have gone crazy (70). This therapeutic power of nature—dirt and flowers—seems to accompany her during her "desert exile" as well. It is therefore significant that, among the few possessions she carries when sent to Poston camp, she chooses to include the stock seeds, which she smuggles out of the farm (78).

When Sumiko arrives in Poston, she finds it difficult to accept her lot, as the other prisoners have done. She misses her life at the farm and, much like Sone and her Idaho nights, she only finds comfort and consolation when "lying outside at night under the stars" (Kadohata 2006, 129). Although she does not find the scent and the serenity that the flower farm afforded her, Sumiko can at least see the same sky that she saw back home. One day, out of boredom, Sumiko starts watching Mr. Moto's preparations of "his soon-to-be garden" (126), a project for which he lacks the flower seeds that the girl has secretly hoarded all this time. This is where we readers are reminded of the girl's close identification with weedflowers, confirmed by the fact that, before leaving the farm, her uncle had called the new strains for stock weedflowers "Sumiko" (78). In giving Mr. Moto her seeds, then, Sumiko is offering her-self; she is trying to survive through those seeds. When Mr. Moto asks her to help in the gardening project, Sumiko feels elated and she soon starts "planning her section of garden" (145). While her expert "diagrams" apparently fit traditional agrarian patterns, Sumiko complements such a vision with a genuine love for less domesticated nature, like wildflowers and dirt itself. In fact, prior to Poston, "dirt" had been Sumiko's "favorite thing" (149), utterly devoid of the negative connotations that would accrue to it in camp, where it would be associated with the nuisance of dust storms (101–102). The girl is aware that while she "*still* loved dirt," it was not the barren dirt of the concentration camp, which "hardly smelled at all" (149). When Sumiko realizes that what she needs is to add organic matter to camp dirt and finally manages to do so, she takes secret pleasure in the sweet scent of real dirt (154). Like weeds and dirt, Sumiko seems negligible and unimportant, but she resists this interpretation both of those natural elements and of her own status. Against all odds and preconceptions, there is agency in dirt (Alaimo 2008, 247), especially in its drive to perpetuate itself and its ability to allow for life forms, like stocks, to emerge.

Despite what the others tell her about the inappropriate weather, Sumiko is adamant in having her flowers grow. Finally, after weeks of patient waiting, she sees the first results: "Three gorgeous tendrils of green poked out from the dirt" (Kadohata 2006, 152). Sumiko then succumbs to "a wave of paranoia" and feels anxious and protective of the fragile plant, a care that arises as much from her previous farming experience as from her ability to feel for the plant.[46] Fearing that "sprouts" like those green tendrils are likely to be "a treasure trove to a desert bug," Sumiko decides to build a little dome out of sheet in order to shield and protect them (152). When the plants finally bloom, Sumiko is beaming and proud of her "flower forest" (153). What proves most revealing in this description of Sumiko's weedflowers, much like Mother's dandelions in Sone's book, is their color. The girl is astonished because, instead of being uniformly peach, the flowers she has planted turn out to be "the colors of the rainbow" (152). The symbolic import of the plants is evident in the context of racist incarceration: the book responds to the racist project of homogeneity by positing the alternative icon of the multicultural rainbow.[47]

The birth of environmentalist attitudes is nowhere more evident than in Miyake's neo-internment novel, *21st Century Manzanar*. Rhetorically speaking, the novel uses and revamps the metaphor of literal incorporation, of "swallowing," introduced in our previous discussion of environment shock, where I argued that most narrators imagined the land itself, in its synecdochic form of mud or dust, invading or entering them, in a disturbing case of transcorporeality.[48] In *Nisei Daughter*, for instance, Monica described the moment of arrival as one of being figuratively "eaten" by the new environment, through its "glutinous mud" (Sone 1953, 173). In a more oblique way, *Farewell to Manzanar* also depicted the landscape "swallowing" people. When Jeanne enumerated the pictures included in the "high school yearbook for 1943–1944" (102), she described some in more detail, in particular some of Miyatake's "unofficial" photographs:

> finally, two large blowups, the first of a high tower with a searchlight against a Sierra backdrop, the next a two-page endsheet, showing a wide path that curves among rows of elm trees. White stones border the path. Two dogs are following an old woman in gardening clothes as she strolls along. She is in the middle distance, small beneath the trees, beneath the snowy peaks. It is winter. All the elms are bare. The scene is *both stark and comforting*. This path leads toward one edge of camp, but the wire is out of sight, or out of focus. The tiny woman seems very much at ease. She and her tiny dogs seem almost *swallowed* by the landscape, or *floating* in it. (Houston and Houston 1973, 103; emphasis added)

While Sone's rhetorical use of the invading mud had clearly negative connotations, Houston's description of this image is much more ambivalent, "both stark and comforting." The woman in the photograph is perceived as comparatively "small," or "tiny," dwarfed as she is by the tall trees and the even taller mountains in the distance. The natural environment seems to be the protagonist, not the barbed wire—which the narrator cannot see—nor the woman and her animal friends. The apparent insignificance of the woman, however, does not detract from her enjoyment and peace, since she is clearly "at ease" in that impressive landscape. The new environment may seem to "ingest" her, but the alternative verb "floating" echoes the initial ambiguity (stark/comforting) of the scene.[49]

Trans-corporeal metaphors of ingestion and interpenetration are used in a more striking manner in *21st Century Manzanar*, where Miyake deploys trans-corporeality as a positive strategy. Surreal or "spiritual" visions of nature start to proliferate toward the end of the book, in chapter 57, after the protagonists have escaped the camp and found haven in a Navajo/Dineh reservation. Narrated through variable focalization, this chapter first describes the Japanese American refugees, along with their closest friends, making the ascent to the hills, to a Native "medicine wheel" (Miyake 2002, 352). In the following pages, thanks to shifts in focalization, we see how each of the members of this wheel, even the most skeptic ones, manages to recognize their loved ones in the landscape or else experiences a vision of merging with nonhuman nature: Carlos looked around him and "was surprised to discover that the world was not gray" (356); the skeptical Dr. Sid cannot understand "why that mountain was smiling at her" or "why, the longer she stared at it, the more it looked like her brother," who had been killed at Manzanar (357); Greg heard jazz music in the wind and saw the faces of his many dead relatives in the ridges of the cliff (358). David, meanwhile, "saw animals locked in futile embrace. Frozen in death inflicted upon each other," while he witnessed "trees … coming to life" (362).[50] However, it is during his friend Vic's "turn" at focalizing the scene when, once more, the rhetorical use of incorporation or "swallowing" reappears, albeit in a modified manner. While sharing the pipe at the medicine wheel, Vic felt herself flying "way up into the air until she could see the tops of the mountains," "eye to eye with a hawk," and just as she was going "to hit the ground, she'd keep on going, *melding* with the ground, becoming part of the red rock. Her feet and legs would sink into the earth, but *she didn't feel like she was being swallowed up*," but rather as if she were "becoming part of the rock," "sinking deeper and deeper until she was looking up at the wheel from below" (358–359). Here, Sone's negative perception and Houston's ambivalence disappear and, instead, we have a genuine bond with the landscape that does not involve annihilation, disappearing into the land, but being part of it, fully embracing trans-corporeality. The belief in a deeply interrelated and interwoven biotic community underlies this last vision, and it seems to be the one that the novel ultimately subscribes to.

Although, prior to the prison escape, Miyake's narrative did not seem to endorse this ecological stance, a more attentive reading of *21st Century Manzanar* unearths earlier signs that prove rather telling, most notably the scenes where human ashes and desert dust seem to blend. When Kate's husband and lover both die in camp, she decides to scatter their ashes in the Manzanar desert, beyond the fence. As she is trying to perform this personal ceremony, she is buffeted by the dusty winds of a sudden sand-storm: "Kate fell to her knees and crawled to the fence, pushing the two tins in front of her as the brutal winds whipsawed through her thin cotton dress and stung her back" (230). After saying goodbye to her lover and letting his ashes fly away, she now tries to do the same with her husband Ray's ashes, but the dust storm makes it more and more difficult: "the wind pushed all thoughts from her head" (232). As she is about to give up, she remembers Ray's peculiar gait and tries once more:

> She pried open the lid and looked up.
> The fence loomed high in front of her.
> The wind whipped the handkerchief from her face.
> She stood up, arms wrapped tightly around the can, and steadied herself, leaning back as a huge gust nearly lifted her up into the barbed wire.
> She saw him walking. Walking away from her.
> She lifted the lid and held the canister over her head.
> The ashes blew away from her in a cloud.
> …
> *Ray's ashes sailed through the barbed wire fence, unimpeded.*
> As if there was *no fence* at all.
> …
> The white cloud of Ray's ashes *sailed across the desert*, riding the devil wind, away from the mountains, *away from the barbed wire fence*, out toward miles and miles of empty, *unimprisoned desert*. (232–233; emphasis added)

Both the lexical choice and the pattern of repetition in this scene emphasize the idea of freedom: the desert is indirectly transformed into an open ocean in which to sail, the barbed wire fence is repeatedly left behind, the participial adjectives *unimpeded* and *unimprisoned* point at the end of obstacles and confinement. At the same time, the ashes fly away from the camp but remain in the desert, a desert now devoid of the bleak connotations that had initially accrued to it.[51]

A very similar scene takes place in the Dineh reservation, after the main characters flee the concentration camp. Once they feel safe, Kate and David decide to say goodbye to their loved ones, this time David's friend Christine and Kate's own son, Graham, who have both died in camp. When Kate opens the canister with her son's ashes, they mix with the canyon that their Dineh friends consider sacred: "Graham's ashes seemed to move in a cloud straight out from the cliff then dip down into the canyon, flitting above the tips of the trees grown up from the bottom of the valley below. They disappeared into the highest branches of the trees as a hawk flew up into the sun" (365). Then David and his Dineh friend, Rodney, do the same with Christine's ashes, which fly "away from them, into the earth, into the trees, into the sky" (365).[52] By voluntarily mixing or blending with the desertic land around them, the prisoners have become "honorary" natives to the land they had been forced to inhabit (the camp desert) and take refuge in (the Dineh reservation desert). As their Native friends remind them, "[i]f you decide to stay [in the reservation] you must realize that you will change and we will change. Dineh strength is that we adapt and survive, as you have. ... You may not be part of this country, but you are part of this land" (375). Like Wong's exotic plants turned natives, discussed in Chap. 4, these "outsiders" have adapted to, adopted, and been adopted by the new land. The ultimate sign of this mutual adoption is the trans-corporeality inaugurated by this voluntary material intermixing.

3   The Pastoral and the Sublime: Escapism or Empowerment?

Reading these internment narratives from an ecocritical point of view makes us realize that most of the scenes describing the pleasure derived from nature are couched in terms that evoke two key concepts in environmental criticism: the pastoral and the sublime. While some of these descriptions, by foregrounding astonishing vastness and magnificence, seem indebted to Burke's notion of the sublime, others can be more fruitfully analyzed as examples of the pastoral mode, with its idealization of nature and with what many critics continue to see as its escapist connotations. The two visions of nature can even coexist in the same scene, as illustrated by Monica's portrayal of the arid environment in *Nisei Daughter*, which wavers between the calm, "friendly" beauty of the Idaho nights and a landscape that seems to her "strange" and "fierce" in daylight (Sone 1953, 193). While the former description seems more germane to the pastoral discourse, the latter closely approximates the Burkean sublime.[53] In what follows, I will attempt to tease out the multiple ideological strands that underlie these two visions of nature.

In his treatise *A Philosophical Enquiry into the Origin of Our Ideas of the Sublime and Beautiful*, published in 1757, Edmund Burke compares the traditional understanding of beauty as pleasing loveliness with the concept of the sublime, which, according to him, conjures up fear and awe rather than harmonious aesthetic pleasure. After describing the causes and effects of the sublime and beautiful, Burke concludes that these two notions are antithetical. While the sublime "has terror for its basis" and "causes that emotion in the mind" which Burke describes as "astonishment," the concept of the beautiful is based on "mere positive pleasure" and provokes "love" rather than wonder. Therefore, the feelings resulting from the sublime appear to be much more intense than those triggered by the contemplation of mere beauty; not only that, but the deep response prompted by the sublime can be summoned to enact some change in the viewer. For some, however, the subversive potential of Burke's sublime would in time be co-opted. In *Uncommon Ground: Rethinking the Human Place in Nature* (1995), William Cronon links the development of the cultural construct of the sublime with that of wilderness and explains how the romantic sublime of the late eighteenth and early nineteenth century came to be tamed:

> By the second half of the nineteenth century, the terrible awe that Wordsworth and Thoreau regarded as the appropriately pious stance to adopt in the presence of their mountaintop God was giving way to a much more comfortable, almost sentimental demeanor. As more and more tourists sought out the wilderness as a spectacle to be looked at and enjoyed for its great beauty, the sublime in effect became domesticated. (75)

Still, even if we construe this erosion or co-optation as temporary, not final, the concept of the sublime continues to pose problems.

Let us now go back to Burke's essay, to the passage in which he compares the beautiful and the sublime by contrasting the feelings each elicits in us: "There is a wide difference between admiration and love. The sublime, which is the cause of the former, always dwells on great objects, and terrible; the latter on small ones, and pleasing; we submit to what we admire, but we love what submits to us; in one case we are forced, in the other we are flattered, into compliance" (Burke 1757). If we were to translate this distinction to the realm of the appreciation of nature, we would end up facing a dilemma: on the one hand, a vision of the natural environment as "beautiful" would entail nature's "compliance" and subservience to the human eye/I, a vision rather germane to an extreme form of anthropocentrism that can have dangerous consequences; on the other hand, approaching the environment as a "sublime" element would annul human agency at the expense of awe-inspiring and terror-inducing nature, ultimately taking us to the radical position of deep ecology, at the other end of the environmentalist spectrum.

The concept of the pastoral is not free from controversy, either. By the end of the twentieth century, "pastoral" was "not only a 'contested term,' but a deeply suspect one" (Gifford 1999, 147). More often than not, the pastoral was understood as a "retreat into the past, into nostalgia and often into the country, physically or symbolically" (Helphand 1997, 119; cf. Ladino 2012). This constituted a perilous movement, in that what was left behind was entirely erased: out of sight, out of mind. As a pejorative label, therefore, the word "pastoral" was attached to those texts that dangerously idealized rural life, often by eliding the sociopolitical context behind it, forgetting the stark realities of being a farmer or a shepherd, or disregarding urgent environmental concerns (Gifford 1999, 1–5). However, pastoral discourse does not necessarily prompt a univocal reading. As we saw in Chap. 4, pastoralism can be either "complex" or "sentimental" (Marx 1964, 25); it can be construed as either "oppositional or escapist" (Gifford 1999, 44). In what follows, I will try to ascertain the manner in which the versatile discourse of pastoralism has been mirrored, appropriated, or subverted in internment literature and whether the concept of the sublime has been employed in these internment narratives in perilous or productive ways.

An obvious difference between Japanese American internment literature and the classic pastoral genre is that the retreat into the "countryside" is not voluntary but forced. In addition, the possibility of a return to the "court" is not always clear. In Miyake's novel, for instance, David wishes to escape not so much from the city—although he is forced to leave LA—as from the camp. Having said that, similar situations of exile into a remote, usually sparsely populated or deserted location have been read as complex forms of pastoralism. Bearing this in mind, it is nonetheless fruitful to approach internment narratives as occasionally endorsing, but more often questioning the discourse of the pastoral.

Both fiction and non-fiction about the Japanese American incarceration show that the prisoners, despite the initial feeling of displacement and environment shock analyzed earlier, derived comfort from their active contact with nature. Scholars and writers agree that gardening and landscaping activities helped assuage the traumatic experience of the concentration camps. Those activities, Tamura explains, "buffered the psychological and physical trauma of the incarceration experience" (2004, 2), to the extent that such creative interaction with their natural environment could then be construed as "horticultural therapy" (11), akin to the "spiritual resistance" exercised by people in European ghettos at the same time

(Helphand 2006, 198; see Chiang 2010, 250). Thus, the Japanese Americans' involvement with nature seemed to enable them to forget or at least momentarily "tune out" their hard living conditions and their fundamental deprivation of freedom.

Notwithstanding its therapeutic power, aesthetic pleasure may become anesthetic in the long run. While the pastoral retreat into nature provided the camp prisoners with the psychological respite they needed, it also tempted them to accept their lot in a resigned manner, as captured in the famous motto *shikata ga nai* or "it cannot be helped." The prisoners' passive contemplation of natural beauty might deflate any urge to rebel against their not so "natural" circumstances. Arguably, there is a sedative, paralyzing element in a purely contemplative stance toward nature, redolent of the passive escapism long associated with the classic pastoral genre. The effects of this "false ideology" (Gifford 1999, 7) in our interpretation of camp gardens are evident: a bland view of the pastoral could sabotage the apparently beneficial effects of working in and with the natural environment. Perceived in this way, then, the incarceration garden could function as "a trite palliative, a screen, cover or mask," and "working in the garden" could be (mis)construed as "an abdication of involvement and responsibility" (Helphand 1997, 119). While the internment narratives under analysis offer pastoral moments that may be read in this light, they also include more complex scenes that invite a rather different interpretation.

In Houston's *Farewell to Manzanar*, the father's contemplation of the Sierra mountains that constituted the natural backdrop of the concentration camp can be construed as having the paralyzing effect described above. Watching Mount Whitney, which "reminded Papa of Fujiyama," offered him some solace or "spiritual sustenance," at the same time that it constituted a temptation to accept his situation in a resigned manner.

> He loved to sketch the mountains. If anything made that country habitable it was the mountains themselves, purple when the sun dropped and so sharply etched in the morning light the granite dazzled almost more than the bright snow lacing it. ... They were important for all of us, but especially for the Issei. Whitney reminded Papa of Fujiyama, that is, it gave him the same kind of spiritual sustenance. The tremendous beauty of those peaks was inspirational, as so many natural forms are to the Japanese (the rocks outside our doorway could be those mountains in miniature). They also represented those forces in nature, those powerful and inevitable forces that cannot to be resisted, reminding a man that sometimes *he must simply endure that which cannot be changed.* (98)

Not only here, but elsewhere in Houston's autobiographical narrative, the invitation to forget your plight or resign yourself to the new circumstances is problematically associated either with the contemplation of sublime nature or with the enactment of the pastoral ideal in gardening and agricultural work. As Houston herself recounts, some time after the crisis over the loyalty questionnaire, the Japanese Americans in Manzanar resigned themselves to their lot: "*Shikata ga nai* again became the motto, but under altered circumstances. What had to be endured was the climate, the confinement, the steady crumbling away of family life. But the camp itself had been made livable" (98).[54] It is right after this statement that the narrator describes the ubiquitous gardens and the parks where prisoners like Jeanne herself could momentarily forget their incarceration:

> Near Block 28 some of the men who had been professional gardeners built a small park, with mossy nooks, ponds, waterfalls and curved wooden bridges. Sometimes in the evenings we could walk down the raked gravel paths. You could face away from the barracks, look past a tiny rapids toward the darkening mountains, and *for a while not be a prisoner* at all. You could hang suspended in some odd, almost lovely land you could not escape from *yet almost didn't want to leave.* (99; emphasis added)

Jeanne seems to have been aware of the ambivalent functions of "lovely" nature, at once a welcome balm and an alienating sedative. In addition to endowing nature with both assuaging and escapist roles, the way this scene is described manages to erase not only the signs of imprisonment, such as the barbed fences, but also the prisoners themselves.

The depoliticization of nature, or rather its manipulative use as an instrument of numbing alienation, reappears in recent internment narratives like *21st Century Manzanar* or *Weedflower.* Kadohata's novel alludes to the prisoners' gradual acceptance of their incarceration, especially noticeable in their excursions. As Sumiko recounts in *Weedflower*, after a while adults had so internalized their status as prisoners that even when they were allowed to walk to the mountains unescorted, "nobody ever tried to escape" (Kadohata 2006, 126). This does not seem to be the case in Miyake's *21st Century Manzanar*, which ends with a final escape or "flight into the desert." However, prior to this moment, some characters in the novel had tried to reproduce the pastoral discourse, while others had fallen prey to its temptation of passive retreat. In particular, the sections of Miyake's novel where Lillian becomes the focalizer conjure a preposterous version of the pastoral ideal, where the camp "residents" are the sheep and she becomes the shepherdess of the flock. Lillian not only paints a false, idealized picture of happy campers, but also reaffirms the "anesthetic" role of gardening:

> This was a happy temporary relocation center that would be self-sufficient by the end of summer. The labor-reward system awarding meal chips for each task performed was working splendidly. ... Plus, as long as the water supply held up, there would be a *garden providing serenity* and fresh vegetables, which was important in Japanese culture. (147–148; emphasis added)

As illustrated by this quotation, Lillian's thoughts and utterances provide Miyake with excellent opportunities to denounce the incarceration through irony and sarcasm. However, it is not just the words uttered by the too villainous villain in the novel that reveal the escapist underside of the pastoral regime. The protagonist, David, who clearly resists and resents the incarceration, is also lured by the possibility of forgetting their common plight, their lack of freedom, by focusing on the impressive landscape: "David stared out through the barbed wire fence at the snow-covered mountains looming over the camp. The clouds seemed magnetically drawn to the mountains, brushing the flat desert camp with a brief dusting of snow like an afterthought" (162). During this contemplative interlude, David concedes that he would easily get used to life in Manzanar because of the quiet and the hypnotic natural beauty. In this camp he does not have to drive, he does not have to deal with LA "traffic, gas prices, the freeways, the shootings, the daily incidents of rudeness. He didn't miss those at all" (162–163). Once more, it is by admiring the mountains surrounding him that the prisoner seems tempted to accept his lot and stay in Manzanar:

> David stared out at the white dusted mountains. They didn't move and neither did he. They were all he could see of the outside world and that suited him just fine.
> He had his family with him ....
> He looked out at the stars in the clear, desert sky and the moon laying a blue film on the snowcapped mountains. He breathed in the clean, clear, searingly cold, winter air and let it sting his lungs ....
> The thought of *staying like this forever* flickered across his mind, *comfortingly*. (163; emphasis added)

This scene, redolent as it is of the pastoral topos *locus amoenus*, is rather unexpected in an otherwise abrasive type of narrative. David's daydreaming of pastoral bliss, however, is suddenly interrupted by gun threat.[55]

> "Freeze!" a voice cracked behind him.
> David pulled his gaze away from the blue snow on the mountaintop and turned to see a young white boy in an army uniform pointing an M-16 at David's chest.
> ... "What are you doing so close to the fence?"
> David thought a moment then said, "I was trying to get a view less obscured by the barbed wire."
> "Why?"
> "So I could look at the whole mountain"
> "Why?"
> "So I wouldn't have to see the barbed wire." (163)

The rapid dialogue between David and the soldier once more highlights the ludicrous nature of the situation, reflected in the circularity and the absurdist tone of the repartee itself, while it also foregrounds what Houston's narrative had erased: the inevitability of the "barbed wire" that constantly reminds internees of their lack of freedom. The fact that David cannot see the "whole mountain" because of the interposed barbed fence meaningfully links freedom with full enjoyment of nature. Thus, David's brief pastoral dream vanishes as the desert mirage it actually was. Moments later, when the soldier finally allows him to leave, David's reflections turn bitter. He realizes he has been duped into believing that life was bearable in a concentration camp, that this was better than having freedom in a dangerous world. "Things hadn't changed" that much after all, David muses about "camp life"; once more, as in LA, he found himself "ducking punks with guns" (164).

However, not all visions of the natural environment fit Lillian's bland pastoral dream of "serenity-providing" landscapes. Several scenes in these internment narratives are indebted to the discourse of the sublime, with its emphasis on an overwhelming feeling of awe. What sublime art or nature provokes in us is not uncomplicated pleasure, Burke (1757) reminds us, but a specific type of "delight," which he describes resorting to an apparent oxymoron: "a sort of delightful horror, a sort of tranquillity tinged with terror." In the internment narratives under analysis, the awe-inspiring sublime is most noticeable in the mixture of fear and fascination that some characters feel when they first encounter the desert vacuum in its terrible immensity. As Burke explains when enumerating the characteristics of the sublime, in comparison with beautiful objects, "sublime objects are vast in their dimensions." This is precisely the feature that David highlights when he first lays eyes on Manzanar: he is "blown away by the *immense*, barren *vastness* engulfing them" (Miyake 2002, 91; emphasis added). Not only are the huge dimensions underlined, but the character's initial reaction clearly echoes what Burke considers the primary effect of the sublime: astonishment. Nevertheless, as we saw in our earlier analysis of Miyake's brown sedan scene, the shock produced by the characters' encounter with the sublime has a negative effect: that of paralyzing the viewers and rendering them less alert to the very real dangers that surround them.

Despite the negative connotations that have traditionally accrued to (most uses of) the pastoral and the sublime, both discourses have been recently redefined along more subversive, less compliant lines. Firstly, the "discourse of retreat" that lies at the core of the pastoral genre can be put to two radically opposite uses: it can be deployed in order to "*escape* from the complexities of the city, the court, the present," or it can be used precisely to *explore* and ponder on those same complexities (Gifford 1999, 46). Not only that, but the classic pastoral begets a new form of pastoral discourse that can better be captured by the term post-pastoral. First explored by Gifford in *Green Voices* (1995), post-pastoralism eschews the two dangers that anti-pastoral scholars had criticized in the pastoral tradition: the pitfalls of both neglecting "social reality" and ignoring "ecological concern[s]" (Gifford 1999, 2). Through the post-pastoral mode, literary texts both continue the tradition of the pastoral, with its emphasis on the retreat-return dialectics, and critique its mystifying uses in the best tradition of anti-pastoralism.

In a similar fashion, a few environmental scholars like Helphand (1997) or Outka (2008) have drawn attention to the liberatory power of Burke's discourse of the sublime. In *Race and Nature from Transcendentalism to the Harlem Renaissance*, Outka discusses the different traditions of the sublime, including Kant and Burke's theorizations, and he praises the anti-racist potential behind the "ecological sublime" (2008, 201–203). Likewise, in his exploration of defiant gardens, Helphand does not link this "domesticated" or middle landscape with the pastoral, as readers might expect, but with the concept of the sublime. It is Burke's concept that supports Helphand's "defiant garden esthetic," a vision that is not so much about "escape" as about "involvement": "The garden can be assertive as well as passive"; it does not only conjure up "a calm retreat and welcome respite," like the classical pastoral, "but also the assertion of a proposed condition, a model of a way of being and of physical form" (Helphand 1997, 119). The defiant gardens that Helphand sees in concentration camps would then be heirs to the best aspects of Burke's notion of the sublime, including its potential for astonishment.

Nevertheless, it could be argued that the defiance implicit in the internment camp gardens was oblique and moderate, rather than overt and radical. Thus, by creating such "defiant gardens," Japanese Americans would be treading a middle ground between absolute submission and open resistance. However, there continue to be solid arguments in favor of the "defiant" nature of gardens. In fact, if we approach gardens and landscaped parks as sites for thinking and preparing action, such domesticated landscapes would then have a huge subversive potential. Read along these lines, those gardens would constitute not so much the locus of nostalgic alienation as strategic sites allowing prisoners to recover their sense of human dignity, a necessary preliminary stage that would prepare them for resistance. What is more, the association of gardens with a locus of resistance is not exclusively metaphorical: it could be literal. In Manzanar, the elaborate garden adjacent to Block 22 not only played a healing aesthetic role, it also constituted "a place of covert and overt defiance, vehement protest, and then the organizational center of the Manzanar riots" (Tamura 2004, 17) that shook the camp in December 1942.

If gardens could become both landscapes and mindscapes for action, the excursions into the camps' surrounding nature equally harbored an immense liberatory potential. These outings, although initially organized—and apparently co-opted—by the authorities, gave the prisoners a taste of freedom that was difficult to forget. Even if the purported aim of supervised excursions was to preempt Japanese American resistance to incarceration, the hiking trips to the desert and its oases constituted a temporal and ideological gap in the incarceration discourse, a moment in which the inmates could also develop a free and beneficial relationship with the land.

Even within the anthropocentric framework of human dominion over nature, activities like gardening, landscaping, and farming proved highly subversive in the context of incarceration, in so far as such activities literally claimed the new natural environment that Japanese Americans had been forced to inhabit. By gardening and farming the land that had been imposed on them, Japanese American prisoners symbolically claimed that territory as theirs. Thus, although the choice of creating gardens near their living quarters was initially of a practical nature, such a choice was later endowed with political symbolism. Helphand sees this garden making as an evident locus of resistance: the "first act" of defiance among these prisoners "was to *appropriate* land without permission" (Helphand 2006, 198, emphasis added).[56] However, this discourse of "appropriation" is not without its perils, as explored in Chap. 4, since it seems to endorse settler colonialism, as well as capitalist attitudes that view the land as mere property.

Much more rewarding and valuable for our present discussion is the horizon of possibilities opened by the post-pastoral mode. In the last part of *Pastoral*, Gifford describes the six characteristics encountered in post-pastoral texts: first, the humbling of humanity; second, the recognition of the creative and destructive sides of nature; third, the connection of inner human nature and external nature; fourth, the newly acquired knowledge that culture and nature are mutually imbricated with each other; fifth, the movement from consciousness to conscience, from awareness to an environmental ethics; and sixth and last, the ecofeminist insight that explicitly links environmental exploitation with other forms of oppression and discrimination, most notably sexism and racism (Gifford 1999, 152–169).[57] Arguably, another post-pastoral characteristic, the recognition of our intimate, material connection with the external, has also come to the fore in recent years, especially with the emergence of material ecocriticism and, in particular, the concept of trans-corporeality. The last chapters of Miyake's novel seem to embrace all of the above features, which leads us to read this neo-internment narrative as post-pastoral.

4   CULTIVATING THE ANTI-CAMPO

Let us conclude this exercise in literary topography with an etymological note. As we know, Latin *campus*, from which both *camp* and Agamben's *campo* derive, means *field*. The concentration camps erected to confine Japanese Americans during WW2, faithful to their etymological nature, finally became replete with gardens and adjacent farming fields. The correlation between field and *campo* was underscored by the fact that the end of gardening and farming marked the end of the concentration camps, and vice versa: "That summer [of 1945] the farm outside the fence gradually shut down. Cultivation stopped. ... Nothing new was planted. ... [T]hen the word went out that the entire camp would close without fail by December 1" (Houston and Houston 1973, 133). The foregoing analysis of internment narratives proves that the prisoners' farming, landscaping, and gardening activities had a political potential beyond their practical uses. Initially, the internees, resorting to different cultural legacies, seemed to embrace the domesticating ideal: they would tame this strange, "wild" territory they had been "given." This, however, implied a discourse of appropriation that could raise doubts from a conservationist (and decolonial) point of view, yet it did not tell the entire truth about the prisoners' attitudes toward their natural environment. As I have tried to demonstrate, these internment narratives trace the internees' evolution from the initial domesticating patterns of agrarianism to a different discourse, where the main concern is no longer the radical modification of the natural environment but the appreciation of, adaptation to, and even conservation of the ecosystem. More importantly, the prisoners depicted in these texts, initially perceived as Agamben's *homo sacer*, struggle to acquire agency and dignity by first becoming *homo faber/agricola* and then, in the most enlightened cases, *homo ecologicus*. If, as Helphand skillfully puts it, the "defiant gardens" created by Japanese Americans in their concentration camps became the *anticamp* (2006, 189), the defiant narratives under analysis, in their deft handling of the interaction between the human prisoners and nonhuman nature, equally portray the emergence of an *anticampo*. The shift from *homo sacer* to *homo ecologicus* effectively subverted the campo paradigm, eroding the dehumanization and disenfranchisement inherent in it. Deprived of their material possessions, Japanese Americans resorted to the natural environment in order to create something they could call "their own," a sense of "ownership" that evolved from the initial discourse of domestication and possession to a more respectful attitude of cooperation and conservation. Deprived of one of the fundamental human rights, freedom, Japanese Americans still managed to retain some degree of agency by cultivating an ecological anti-campo.

NOTES

1. The history of Japanese Canadians during the war is very similar. In her oral history project on Japanese Canadian internment, Pamela Sugiman explains how, after Pearl Harbor, Japanese Canadians "were displaced from their homes in British Columbia (BC), dispossessed of property, housed temporarily in former horse stables, interned in remote ghost towns, incarcerated in prisoner-of-war camps, and forced to work as low-wage agricultural labour" (2009, 187). There were certainly some differences, especially in the degree of family disruption and in the fact that Japanese Canadians were not allowed to move freely in the country until 1949. Until then, British Columbia was off-limits for people of Japanese ancestry (Sugiman 2009, 192).

2. The estimated number of Japanese Americans who were sent to intern-ment camps wavers between 100,000 and 120,000 depending on sources. In any case, it was the vast majority of the 126,947 Japanese American living in the continental United States by 1940 (Helphand 2006, 155).

3. For an analysis of the role of photography in documenting and (re)con-structing the internment experience, see Gessner (2005), Hesford and Kozol (2001), Graulich (2004), and Phu (2008).

4. As Limerick had done, researchers started to praise internment gardens as "cultural resources evocative of human agency within landscapes of persecu-tion and racism" (Tamura 2004, 1), while "garden making" was construed as "an open and sure sign of internees exerting control over their situation" (Helphand 2006, 166). In *Artifacts of Loss* (2008), Jane Dusselier argued that Japanese Americans purposefully interacted with their environment, turning camps into "survivable places"; through gardening and landscaping the internees "drew on the physical landscape to reshape understandings of power and generate new ideas that better ensured their survival" (51).

5. Sarah Wald devotes one of the chapters of *The Nature of California* to a less-known Japanese American narrative penned during the internment years: Hiroshi Nakamura's *Treadmill*. Discovered in 1952, the novel was not published until 1996 (Wald 2016, 87).

6. I am aware of the dangers involved in a transhistorical analysis of intern-ment literature. There are obvious differences between the older narra-tives, often penned by Nisei writers who had gone through the incarceration themselves, and the more recent "neo-internment" novels written by Sansei and subsequent generations. Although these neo-internment narra-tives continue to explore the ways in which life in the concentration camps shaped the lives of several generations of Japanese Americans, they gener-ally depart from the "testimonial" stance of previous narratives, as some critics have pointed out (Beck 2009, 8; Manzella 2013, 146, 157–8). At the same time, much can be gained from a comparative analysis of books dating from different historical periods: without ignoring the differences outlined above, such a study allows us to see the continuities within intern-ment literature. I thank the reviewers for drawing attention to the conten-tious nature of transhistorical approaches to literature.

7. While environmental historians like Limerick or Helphand have provided an invaluable diagnosis of the internment experience, I maintain that their primarily "factual" studies need to be complemented by literary criticism, for some of the documents that serve as the basis for their sociohistorical interpretation are first and foremost literary texts. Traditionally, literary critics have either ignored the role of the environment in these texts, or else they have interpreted it as just a passive background. In contrast to this critical tradition, in what follows I will argue for the unforeseen relevance of the (un)natural environment in narratives that apparently deal with something entirely different.

8. This section and parts of the analysis of Miyake's novel have appeared, in a modified form, in "Revisiting the Campo: A Biopolitical Reading of Perry Miyake's *21st Century Manzanar*," published in the *Revista de Estudios Norteamericanos* 20 (2016). I thank the journal editors for allowing me to reprint part of that article.

9. To start with, the word *relocation* implied movement rather than permanence, while camp life had a semipermanent ring to it. Moreover, the term *relocation* was employed both for the removal of Japanese Americans from their homes all along the West Coast and for the process of internal migration during the years of incarceration and after the camps were eventually closed (Houston and Houston 1973, 84). For a more detailed discussion of terminological issues, see Daniels (2005) and Simal-González (2016).

10. In "Words Do Matter" (2005), Roger Daniels puts forward a convincing argument for the inaccuracy of the term and advises critics and historians to talk about *incarceration* and "concentration camps" rather than *internment*. To this day, however, the use of "concentration camp" to refer to the Japanese American camps remains polemical because of the negative connotations associated with the phrase. In order to preserve the familiarity of the phrase "Japanese American internment" and at the same time follow Daniels's advice, in the present chapter I will be employing both terms, *internment camp* and *concentration camp*. Such a combined use partially dilutes the genocidal echoes of the latter phrase while also problematizing the neutrality of the former. When referring to the particular literary subgenre, I will be using the phrase "internment literature."

11. For a detailed analysis of Miyake's novel from a biopolitical perspective, see Simal-González (2016).

12. Scholars like Alan Wald or Roger Daniels have pointed out the incommensurability of the treatment of people of Japanese ancestry during WW2. Germans and Italians did undergo internment, but it was in accordance with the Geneva conventions, that is, as enemy aliens and in an apparently legally justified manner (Daniels 2005, 191–192). For Fred Lee, Japanese American concentration camps constitute valid evidence of the "racial state of exception" (2007, para. 21) imposed in the United States during WW2.

13. It is worth noting that the Japanese immigrants in mainland America were demographically and economically marginal in comparison with the Japanese in Hawai'i, and yet the latter did not suffer the experience of collective internment (Sone 1953, 159; see Phu 2008, 338). Only 1440 Japanese Americans were imprisoned in Hawai'i, in comparison with the 110,000 to 120,000 evacuated and sent in concentration camps in Western continental United States (Helphand 2006, 156). For a history of the Japanese Hawai'ians who were selectively imprisoned and their peaceful attempts to offer resistance, see Okawa (2011).

14. Agamben borrows Walter Benjamin's term *bloß Leven* and traces back the origin of the term *nuda vita* to the Aristotelian concept of *zoe*, in contraposition with political *bios* (1995, 1–14). For the Italian philosopher, it is neither legal discourse nor the sacrificial-religious metaphor of the Holocaust that best explains the massive killings, but the biopolitical paradigm of the *homo sacer* (27). In the old Roman order in which we first encounter that figure, the *homo sacer* constituted a paradox: he could not be sacrificed or ritually immolated, but at the same time he could be killed with entire impunity (91).

15. If Hannah Arendt sees concentration camps as the "laboratories" of modern totalitarianism, Agamben reverses the argument: it is the transformation of politics into the space of *nuda vita*, epitomized by the camp, that has made possible the absolute control of totalitarian regimes (1995, 132).

16. "Se l'essenza del campo consiste nella materializzazione dello stato di eccezione e nella conseguente creazione di uno spazio in cui la nuda vita e la norma entrano in unha soglia de indistinzione, dovremo ammettere, allora, che ci troviamo virtualmente in presenza di un campo ogni volta che viene creata una tale struttura, indipendentemente dall'entità dei crimini che vi sono commessi e qualunque ne siano la denominazione e la specifica topografia" (Agamben 1995, 195).

17. The Japanese American concentration camps have recently been interpreted as the perfect illustration of Agamben's theory. Both Lee (2007) and Sokolowski (2009) have used Agamben's paradigm in their respective explorations of the Japanese American internment. For Sokolowski, it is the apparently coincidental juxtaposition of fleeing refugees in Europe and the Japanese American "relocation" in Okubo's *Citizen 13660* that conjures up Agamben's theory about concentration camps (2009, 74). On the other hand, Lee chooses to interpret the handing out of the loyalty questionnaire to imprisoned Nisei under the light of Agamben's paradoxical *homo sacer*.

18. They also mention entertainment and recreational activities like dances or sporting events. For a discussion of such activities, see Colborn-Roxworthy (2007) and Dusselier (2008). For a list of entertainments in Manzanar, see Houston and Houston (1973, 100).

19. "US literature has long depicted farming as a transformative act," claims Wald; therein lies "its appeal as a trope for immigrant and nonwhite authors" (2016, 8).

20. Such temporary sites comprised "nine fairgrounds, two racetracks, a mill site and an exposition center" (Helphand 2006, 157). Although assembly centers were the norm, some Japanese Americans like those living in Terminal Island (Houston's case), were directly removed to the camps. This is the reason why, in her memoirs, Houston depicts the moment when

her family is directly evacuated to the first camp to open, Manzanar, in Owens Valley. Among the four narratives under analysis, then, only Sone's *Nisei Daughter* and Kadohata's *Weedflower* describe in detail the evacuation to the assembly centers.

21. The complaint of animalization also resounded in the concentration camps, where characters like Mama in *Farewell to Manzanar* voiced their despair at having to live like nonhuman animals: "we can't live like this. Animals live like this" (Houston and Houston 1973, 26).

22. The fact that dogs appeared in Japanese American family portraits as early as in 1914, as proved by the Yoshino's photograph in *The American Japanese Problem* (reproduced in Azuma 1994, 15), not only reflects actual affective bonds between humans and nonhumans, but also responds to the logic of cultural assimilation. That is how Graulich (2004) interprets the presence of dogs in family photographs. See also Dusselier (2008, 144–146).

23. This concern for the abandoned animal friends figures prominently in Kadohata's novel. Like many other children, Sumiko and her younger brother, Tak-Tak, are very fond of animals, especially horses and crickets (Kadohata 2006, 21). In fact, Tak-Tak's best friend is Baba, a mare that helps the family in their flower business: he "really adored Baba. Her nose dripped all the time, but that worked out fine because Tak-Tak liked gooey things" (4). However, when Sumiko and her family are told to leave their farm and they cannot sell Baba to anyone, she is old enough to suspect that the future awaiting the mare is rather grim. Not only horses but also smaller animals had to be abandoned, and very few people or institutions seemed to take care of the pets that these families had been forced to leave behind: "Humane societies in evacuation areas were supposedly overflowing. Mrs. Ono hadn't been able to find someone to adopt her dog, and she refused to have the animal put down: instead, she had left several open bags of food in her house and tacked a note about the dog on her front door" (80). Sumiko is sure that "nobody would want a 'Jap' dog" (80), but, surprisingly enough, Mrs. Ono receives a letter from a neighbor who is willing to adopt the animal (93).

24. The difference between pets and pests, already discussed in Chap. 3, is probably the most obvious or extreme case of this hierarchical taxonomy whereby some animal creatures are expendable while others are not.

25. I thank the anonymous reviewers for having encouraged me to tackle these issues and explore not only the bonds formed with nonhuman animals but also why certain animals were kept while others were released. For a more detailed discussion of animal rights and speciesism, see Chap. 1.

26. The phrase "environment shock" encompasses both landscape and climatic aspects. Although some scholars have analyzed climate shock in human and nonhuman species, few have focused on the psychological effects that a radical change in the landscape has upon human beings (the analysis in Pearce 1981, one of the exceptions, is restricted to the impact this shock has on tourists). I favor the phrase "environment shock" over "ecosystem shock," used in studies whose primary focus is not human beings (e.g. White, Murray, and Jackson's 2004 report for the National Wildlife Federation).

27. Although changes in the natural environment were already visible in some of those temporary centers, in general these were located "close to home," along the West Coast, so at least "the geography and climate" remained more or less "familiar" (Teorey 2008, 231). Even when Japanese Americans were allowed to see the landscape as it gradually changed, the specific trajectory of the journey to their permanent camps was kept secret from most evacuees. In fact, window shades remained drawn for most of the trip, especially when the trains or buses went past military or other "sensitive" areas, as well as through "hostile country," as Sone puts it in *Nisei Daughter* (1953, 190–191). Similarly, Kadohata describes how, during the first journey, the military trucks also remained covered: "A soldier closed a canvas flap, and darkness enveloped the back of the truck," where Japanese American evacuees had been placed (2006, 82). In the second journey, from assembly center to concentration camp, much the same thing happened: the shades had to be kept drawn for the entire journey (97–98), apparently for the Japanese Americans' own safety.

28. However, as Limerick cogently elucidates, most Issei and Nisei no longer saw the West Coast as "a newly encountered landscape" but as "home" (1992, 1037, see Chiang 2010, 242).

29. See, for instance, the scenes described by Sone (1953, 192–193), Houston and Houston (1973, 87–89), Miyake (2002, 221, 224, 230–233), or Kadohata (2006, 101–102).

30. Barbara Kingsolver ironically captures this popular perception of deserts and arid landscapes when she describes "a trackless wilderness in western Arizona that most people would call Godforsaken, taking for granted God's preference for loamy topsoil and regular precipitation" (2001, 128).

31. Alaimo's notion is similar to Nancy Tuana's concept of "Viscous Porosity" (2008). Like them, in his 2012 study of how urban environment is depicted in contemporary US ethnic literatures, John B. Gamber focuses on the permeability of boundaries and the "positive pollutions" occurring between them.

32. At first sight, the ethics of conservation is not so obvious in historical records, newspaper articles, and diaries as it is in internment literature. The

former documents suggest that, at the beginning of their incarceration, Japanese American prisoners merely thought of reproducing the natural environment they were familiar with and putting their agricultural skills to good use. However, even in these historical documents, there is a perceivable shift in the internees' attitudes. In the course of the three or four years of their evacuation and internment, they moved from the initial rejection of the imposed landscapes to a respectful adaptation and appreciation of its beauty, by means of an increasing knowledge of the new environment. See Limerick (1992) and Helphand (1997, 2006) for an in-depth study of the camp newspapers and official records.

33. See Daniels et al. (1986), Azuma (1994), Helphand (1997, 115), and Helphand (2006, 157–59).

34. See Tamura (2004, 5–6, 18–19), Helphand (2006, 155), Miyake (128, 211, 315–316), and Kadohata (24). This urban gardening expertise came handy when internees, with the WRA authorities' approval, undertook the task of "beautifying" desert camps like Minidoka, Poston, or Manzanar. For an in-depth study of the gardening and landscaping activities in Minidoka and Manzanar, see Tamura (2004) and Helphand (1997, 2006).

35. In Kadohata's *Weedflower*, the comments by both white and Native American characters describe the extent to which Japanese Americans were perceived as experts in gardening and agriculture: "The Japs sure have a knack for growing things" (Kadohata 2006, 120); "My brother said the Japs are all farmers" (Kadohata 2006, 121); etc.

36. Limerick seems to favor this hypothesis when she argues that, apart from fulfilling practical purposes, gardening and landscaping activities in internment camps were the result of a "powerful set of aesthetic and spiritual motivations, derived directly from Japanese culture" (1992, 1042). Following Dusselier's advice (2008, 9–10), I would qualify Limerick's statement in order to divest it of its essentialist connotations. It may be true that, in their interactions with their new environment, many Issei resorted to Japanese cultural practices even as they adapted them to the new context. Yet the creation of gardens and recreational spaces as a way to claim agency and build a better environment is not unique to the Japanese American incarceration but appears in similar situations of oppression and incarceration like those analyzed in Helphand's *Defiant Gardens* (2006).

37. In an article from the *Topaz Times* the Japanese American writer explicitly places the internees within the paradigm of settler colonialism: "We are again the pioneers, blazing the road into the wilderness of our social frontiers" (quoted in Helphand 2006, 193); accordingly, he interprets the concentration camps as "frontier communities in wild landscapes, which internees 'civilized' with their frontier gardens" (Helphand 2006, 192).

Similarly, Chiang quotes a *Minidoka Irrigator* article where one of the prisoners describes "the formidable, but admirable, task of *conquering* the frontier," "Our great adventure is a '*repetition of the frontier struggle of pioneers against the land and the elements*'" (2010, 248). Unless we understand such references to their role as the "new pioneers" as ironic—and we have no reason to do so—this incongruous use of the American frontier discourse among the prisoners proves rather striking, for it altogether ignores the compulsory nature of the internment experience and seems equally oblivious to the implications of redeploying settler colonial concepts.

38. In fact, a good number of the prisoner characters in these narratives are gardeners and farmers. Even in the apparently least pastoral of these texts, *21st Century Manzanar*, prisoners like Mary's father cultivate spring gardens (Miyake 2002, 111). In Miyake's speculative novel, set sixty years after the first internment—which meant that few Nisei had to go through this second incarceration—the "last Nisei" left in camp is a gardener (128–129), just as David and his (deceased) father used to be (211, 267).

39. Originally an apple orchard, Manzanar had gradually become a desert after a long history of water disputes with the neighboring metropolis, LA (Limerick 1992, 1049; Chiang 2010, 241). For more information about the Owens Valley battles, see Kahrl (1982) and Libecap (2007).

40. "I had a pond garden in my old home," says Mr. Moto, "so I thought I'd do the same thing here, next to my vegetables" (Kadohata 2006, 126).

41. True enough, the evacuees soon created small family gardens. Notwithstanding the small "victories" that individual or family gardens represented, during the first year of relocation larger plans for agricultural "improvement" had come to nothing. As we have seen, most camps were set in remote desert areas; even though watering and tilling these arid lands occasionally yielded good results, more often than not stubborn efforts to make the desert into something else failed.

42. The fear of this new environment went beyond the understandable dislike of sandstorms and extreme climatic conditions. It also suggested the fear of living in a vacuum. The (semi)arid environment in which these camps were set, interestingly enough, is first perceived as a "barren vastness" (Miyake 2002, 91), an empty space full of nothing, or even a non-place (Kadohata 2006, 104). Still, seeing these desert camps as "no-man's land" could also have its advantages. Such ingrained—if not wholly accurate—vision of arid ecosystems have historically led people, from actual pioneers to philosophers and scholars, to construe the desert as a productive void. Thus, the desertic environment has been read as a blank slate that allowed Japanese Americans, stripped of their literal and figurative place in America, to build a new identity (see Chen and Yu 2005, 561).

Sone conjures up a similar picture when the protagonist goes away from Minidoka, leaving her parents behind. On saying goodbye to her parents, the narrator describes the last image she has of them: "They looked like wistful immigrants. I wondered when they would be able to leave their *no-man's land*, pass through the legal barrier and become naturalized citizens. Then I thought, in America, many things are possible" (Sone 1953, 237, emphasis added). The way Sone chooses to describe her parents and their situation bears a striking resemblance to Agamben's *zones d'attentes* (1995, 195), "una localizzazione senza ordinamento," as befits a concentration camp (197). However, in this case, Sone employs this image as an empowering trope: camps are viewed as a productive "no-man's land."

43. There are abundant examples both in internment literature and in oral/historical records that some prisoners "came to recognize the beauty of the environment and became rejuvenated by their brief escapes from the barracks" (Chiang 2010, 247). One of the archival testimonies from Minidoka proves particularly revealing. Arthur Kleinkopf, one of the camp authorities, described how he escorted some Japanese American ladies in their first outing after six months of imprisonment in the camp. He drove them for some time until one of the women asked him to stop the car so that she could step out, saying: "I'd just like to get out of the car, walk over to one of those trees, touch it and put my arms around it"; Kleinkopf acknowledges he could barely fathom "the feelings of the lady who wanted to *caress* the tree" (quoted in Helphand 2006, 190). Although the description may have been aestheticized to suit the readers' tastes, especially by including the powerful image of a "tree-hugger," it remains true that, in these and other scenes, one can glimpse ecological attitudes that were not so common in the 1940s.

44. Interestingly, the new barracks were located next to one of the "few rows of untended pear and apple trees" that had survived the desertification of the 1920s (Houston and Houston 1973, 95).

45. Using this natural trope to signify the Japanese Americans and their plight is a strategy shared by internment narratives ranging from children and young adult novels (Kadohata's *Weedflower* or Tai's *A Place Where Sunflowers Grow*) to speculative novels like *21st Century Manzanar*, most notably when the prison camp is described as a "greenhouse" (Miyake 2002, 30). As we saw in previous chapters, other Asian American texts have made a sociopolitical use of floral imagery. For an in-depth analysis of the topos of natural/social resilience associated with a plant, see my earlier discussion of Shawn Wong's *Homebase* in Chap. 4; for a different use of the floral leitmotif, see the description of Edith Eaton's narrative strategies in Chap. 3.

46. In a rather similar fashion, real Minidoka prisoners "learned to wince when they saw a blade of grass uprooted, a struggling green shoot became a dear thing to be *coddled and petted*. Wistful groups took to gazing longingly at the trees faintly visible beyond the miles on end of sagebrush" (*Minidoka Irrigator*, 1943, quoted in Helphand 179).

47. Although the maneuver may seem relatively facile and predictable, it is nonetheless an effective strategy in a book fundamentally geared at young readers. In the novel, Sumiko also imagines her future in natural terms, as the metamorphosis from chrysalis to butterfly (Kadohata 2006, 115; see Wald 2016, 218–20). This transformation, however, is not so much couched in terms of human rights as in terms of economic opportunities: "from farm girl to flower shop owner" (115–16).

48. When analyzing the "traffic between bodies and natures" (2008, 253), Alaimo illustrates it with the politics of incorporation that often underlie the theoretical foundations of food studies (253–254). However, as she cogently puts it, that "model of incorporation is only one bite away from capitalist consumption" (255).

49. See Graulich (2004, 247–9) for a different analysis of this last photograph, Miyatake's *Woman walking dogs*, as described by Houston.

50. Among them, David sees nonindigenous trees "now threatening to monopolize large amounts of precious water in this desert. Intruders draining the life force of the land" (Miyake 2002, 362). While the environmentalist drift is self-evident in the last quotation, such insistence of non-autochthonous "intruders" is also problematic if read as a sociopolitical metaphor, as discussed in previous chapters.

51. In his introduction to *Getting Over the Color Green* (a title that echoes Wallace Stegner's advice in "Thoughts in a Dry Land"), Scott Slovic explains that it was only in the late twentieth century that American writers started "to recognize the value of vast sprawls of land devoid of human forms and lacking in the verdant, accommodating color green," a color typically privileged in those "cultures that have evolved in the earth's temperate zones" (2001, xvii).

52. This commingling of ashes and desert dust had been anticipated at the beginning of the novel, where Vic suggests scattering her recently deceased brother not in the ocean but in the desert (42). Nonetheless, while Vic, being a Chicana activist, might be thinking of her ancestors' home when she suggested it as her brother's resting place, in the case of the Takedas their choice has to do with symbolic, voluntary emplacement rather than with ancestral (re)claiming.

53. We need to remember that there have been two distinct traditions of the "the sublime" in the West: the classical notion associated with Longinus and his "rhetorical sublime" and the "natural sublime of the eighteenth-

century poets and landscape painters" (Jeffrey 2010, 6), theorized by the British philosopher Edmund Burke. For the purposes of the present study, it is Burke's concept that proves more adequate.

54. It is not so much that Japanese Americans offered no resistance because they were an essentially passive people, as some critics have argued. Their oft-quoted maxim of *shikata ga nai*, translated as "it cannot be helped" (Houston and Houston 1973, 16; Miyake 2002, 265, Kadohata 2006, 130) or "nothing can be done" (Colborn-Roxworthy 2007, 196), can be reinterpreted as the common-sense conclusion after analyzing their situation: what cannot be changed must be endured. Such an apparently fatalist injunction to pull and bear can actually have positive readings as a strategy of resistance much akin to Gerald Vizenor's concept of *survivance*, a combination of "survival" and "resistance." For a discussion of the Japanese concepts of *shikata ga nai* and *gaman* in relation with that of survivance, see Burt (2010) and Shimabukuro (2011). For a definition of the related concepts of *enryo* and *gaman*, see Miyake (2002, 10) and Tamura (2004, 10), respectively. For a more detailed study of *gaman*, see Hirasuna and Heffernan (2005) and Shimabukuro (2011).

55. This is reminiscent of an interrupted idyll or *locus amoenus truncatus* (Simal 2010), a concept to which we will return in Chap. 6.

56. The same discourse of appropriation reappears later in Helphand's analysis, especially when he describes how, on some occasions, "spaces beyond the fences were" not so much given but "*appropriated* as gardens and picnic sites" (Helphand 166; emphasis added). Similarly, Tamura talks of the beautifying "appropriation" of the concentration camps, which "indicated a deep-seated human and cultural need to transform the camp from stranger, symbol of betrayal, and prison into a habitable environment" (7). Through their gardening and landscaping activities, Tamura claims, inmates indirectly "asserted a claim of ownership and the creation of defensible boundaries" (16).

57. We will return to the discourse of post-pastoralism in the following chapter.

REFERENCES

Agamben, Giorgio. 1995. *Homo sacer: Il potere sovrano e la nuda vita*. Torino: Einaudi.

Alaimo, Stacy. 2008. Trans-Corporeal Feminisms and the Ethical Space of Nature. In *Material Feminisms*, ed. Stacy Alaimo and Susan Hekman, 237–263. Bloomington: Indiana University Press.

———. 2010. *Bodily Natures: Science, Environment, and the Material Self*. Bloomington: Indiana University Press.

Azuma, Eiichiro. 1994. Japanese Immigrant Farmers and California Alien Land Laws: A Study of the Walnut Grove Japanese Community. *California History* 73 (1): 14–29.

Beck, John. 2009. *Dirty Wars: Landscape, Power, and Waste in Western American Literature*. Lincoln: University of Nebraska Press.

Burke, Edmund. 1757. *A Philosophical Enquiry into the Origin of our Ideas of the Sublime and Beautiful*. http://www.bartleby.com/24/2/

Burt, Ryan. 2010. Interning America's Colonial History: The Anthologies and Poetry of Lawson Fusao Inada. *MELUS* 35 (3): 105–130.

Chen, Fu-jen, and Su-lin Yu. 2005. Reclaiming the Southwest: A Traumatic Space in the Japanese American Internment Narrative. *Journal of the Southwest* 47 (4): 551–570.

Chiang, Connie Y. 2010. Imprisoned Nature: Toward an Environmental History of the World War II Japanese American Incarceration. *Environmental History* 15: 236–267. https://doi.org/10.1093/envhis/emq033.

Colborn-Roxworthy, Emily. 2007. 'Manzanar, the Eyes of the World Are Upon You': Performance and Archival Ambivalence at a Japanese American Internment Camp. *Theatre Journal* 59 (2): 189–214.

Cronon, William. 1995. The Trouble with Wilderness; or, Getting Back to the Wrong Nature. In *Uncommon Ground: Rethinking the Human Place in Nature*, ed. William Cronon, 69–90. New York: Norton.

Daniels, Roger. 2005. Words do Matter: A Note on Inappropriate Terminology and the Incarceration of the Japanese Americans. In *Nikkei in the Pacific Northwest: Japanese Americans and Japanese Canadians in the Twentieth Century*, ed. Louis Fiset and Gail Nomura, 190–214. Seattle: University of Washington Press.

Daniels, Roger, Sandra C. Taylor, and Harry H. L. Kitano, eds. (1986) 1991. *Japanese Americans: From Relocation to Redress*. Salt Lake City: University of Utah Press.

Dusselier, Jane E. 2008. *Artifacts of Loss: Crafting Survival in Japanese American Concentration Camps*. New Brunswick: Rutgers University Press.

Fiskio, Janet. 2012. Unsettling Ecocriticism: Rethinking Agrarianism, Place, and Citizenship. *American Literature* 84 (2): 301–325. https://doi.org/10.1215/00029831-1587359.

Fujitani, Takashi. 2007. Right to Kill, Right to Make Live: Koreans as Japanese and Japanese as Americans During WWII. *Representations* 99 (Summer): 13–39.

Gamber, John B. 2012. *Positive Pollutions and Cultural Toxins: Waste and Contamination in Contemporary U.S. Ethnic Literatures*. Lincoln: University of Nebraska Press.

Gessner, David. 2005. The Ecotone Interview with Bill McKibben. *Ecotone* 2 (2): 54–59. https://doi.org/10.1353/ect.2007.0062.

Gifford, Terry. 1999. *Pastoral*. London: Routledge.

Govan, Thomas P. 1964. Agrarian and Agrarianism: A Study in the Use and Abuse of Words. *The Journal of Southern History* 30 (1): 35–47. JSTOR. http://www.jstor.org/stable/2205372

Graulich, Melody. 2004. 'Cameras and Photographs Were Not Permitted in the Camps': Photographic Documentation and Distortion in Japanese American Internment Narratives. In *True West*, ed. Nathaniel Lewis, 222–253. Lincoln: University of Nebraska Press. Project MUSE.

Hayashi, Robert T. 2007. *Haunted by Waters: A Journey Through Race and Place in the American West*. Iowa City: University of Iowa Press.

Helphand, Kenneth. 1997. Defiant Gardens. *The Journal of Garden History* 17 (2): 101–121. https://doi.org/10.1080/01445170.1997.10412542.

———. 2006. *Defiant Gardens: Making Gardens in Wartime*. San Antonio: Trinity University Press.

Hesford, Wendy, and Wendy Kozol. 2001. Relocating Citizenship in Photographs of Japanese Americans During World War II. In *Haunting Violations: Feminist Criticism and the Crisis of the "Real,"*, ed. W. Hesford and W. Kozol, 217–250. Urbana: University of Illinois Press.

Hirasuna, Delphine, and Terry Heffernan. 2005. *The Art of Gaman: Arts and Crafts from the Japanese American Internment Camps, 1942–1946*. Berkeley: Ten Speed Press.

Houston, Jeanne Wakatsuki, and James Houston. 1973. *Farewell to Manzanar*. New York: Bantam Books.

Hsu, Hsuan L. 2011. Fatal Contiguities: Metonymy and Environmental Justice. *New Literary History* 42: 147–168.

Jeffrey, David Lyle. 2010. Sacred Proposals and the Spiritual Sublime. In *Through a Glass Darkly: Suffering, the Sacred, and the Sublime in Literature*, ed. Holly Faith Nelson, Lynn R. Szabo, and Jens Zimmermann, 3–21. Waterloo: Wilfrid Laurier University Press.

Kadohata, Cynthia. 2006. *Weedflower*. New York: Atheneum Books (Kindle 2008).

Kahrl, William L. 1982. *Water and Power: The Conflict over Los Angeles Water Supply in the Owens Valley*. Berkeley: University of California Press.

Kingsolver, Barbara. 2001. High Tide in Tucson. In *Getting Over the Color Green: Contemporary Environmental Literature of the Southwest*, ed. Scott Slovic, 128–130. Tucson: University of Arizona Press.

Ladino, Jennifer K. 2012. *Reclaiming Nostalgia: Longing for Nature in American Literature*. Charlottesville: University of Virginia Pres.

Lee, Fred I. 2007. The Japanese Internment and the Racial State of Exception. *Theory & Event* 10 (1). Project MUSE. https://doi.org/10.1353/tae.2007.0043.

Libecap, Gary D. 2007. *Owens Valley Revisited: A Reassessment of the West's First Great Water Transfer*. Stanford: Stanford University Press.

Limerick, Patricia Nelson. 1992. Disorientation and Reorientation: The American Landscape Discovered From the West. *Journal of American History* 79: 1021–1049.

Manzella, Abigail. 2013. Disorientation in Julie Otsuka's *When the Emperor Was Divine*: The Imprisoned Spaces of Japanese Americans During World War II. In *Race and Displacement: Nation, Migration, and Identity in the Twenty-First Century*, ed. Maha Marouan et al., 143–161. Tuscaloosa: University of Alabama Press.

Marx, Leo. 1964. *The Machine in the Garden: Technology and the Pastoral Ideal in America*. London/New York: Oxford University Press.

Miyake, Perry. 2002. *21st Century Manzanar*. Los Angeles: Really Great Books.

O'Brien, David, and J. Stephen Fugita. 1991. *The Japanese American Experience*. Bloomington: University of Indiana Press.

Okawa, Gail Y. 2011. Putting Their Lives on the Line: Personal Narrative as Political Discourse Among Japanese Petitioners in American World War II Internment. *College English* 74 (1): 50–68.

Okubo, Miné. (1946) 1983. *Citizen 13660*. Seattle: University of Washington Press.

Outka, Paul. 2008. *Race and Nature from Transcendentalism to the Harlem Renaissance*. New York: Palgrave Macmillan.

Pearce, Philip L. 1981. 'Environment Shock': A Study of Tourists' Reactions to Two Tropical Islands. *Journal of Applied Social Psychology* 11 (3): 268–280. https://doi.org/10.1111/j.1559-1816.1981.tb00744.x.

Phu, Thy. 2008. The Spaces of Human Confinement: Manzanar Photography and Landscape Ideology. *Journal of Asian American Studies* 11 (3): 337–371. https://doi.org/10.1353/jaas.0.0020.

Rowe, Peter G. 1989. Modern Pastoralism and the Middle Landscape. *Oz* 11: 4–9. https://doi.org/10.4148/2378-5853.1171.

Shimabukuro, Mira. 2011. 'Me Inwardly, Before I Dared': Japanese Americans Writing-to-Gaman. *College English* 73 (6): 649–671.

Simal, Begoña. 2009. Of a Magical Nature: The Environmental Unconscious. In *Uncertain Mirrors: Magical Realism in US Ethnic Literatures*, ed. Jesús Benito, Ana Manzanas, and Begoña Simal, 181–224. New York: Rodopi.

———. 2010. 'The Junkyard in the Jungle': Transnational, Transnatural Nature in Karen Tei Yamashita's Through the Arc of the Rain Forest. *JTAS: Journal of Transnational American Studies* 2 (1): 1–25.

Simal-González, Begoña. 2016. Revisiting the Campo: A Biopolitical Reading of Perry Miyake's 21st Century Manzanar. *Revista de Estudios Norteamericanos* 20: 159–180.

Slovic, Scott, ed. 2001. *Getting Over the Color Green: Contemporary Environmental Literature of the Southwest*. Tucson: University of Arizona Press.

Sokolowski, Jeanne. 2009. Internment and Post-War Japanese American Literature: Toward a Theory of Divine Citizenship. *MELUS* 34 (1): 69–93.

Sone, Monica. (1953) 1979. *Nisei Daughter*. Seattle: University of Washington Press.

Sugiman, Pamela. 2009. 'Life Is Sweet': Vulnerability and Composure in the Wartime Narratives of Japanese Canadians. *Journal of Canadian Literature* 43 (1): 186–218.

Tamura, Anna Hosticka. 2004. Gardens Below the Watchtower: Gardens and Meaning in World War II Japanese American Incarceration Camps. *Landscape Journal* 23 (1): 1–21.

Teorey, Matthew. 2008. Untangling Barbed Wire Attitudes: Internment Literature for Young Adults. *Children's Literature Association Quarterly* 33 (3): 227–245.

Tuana, Nancy. 2008. Viscous Porosity: Witnessing Katrina. In *Material Feminisms*, ed. Stacy Alaimo and Susan Hekman, 188–213. Bloomington: Indiana University Press.

Uchida, Yoshiko. (1982) 2000. *Desert Exile: The Uprooting of a Japanese-American Family*. Seattle: University of Washington Press.

Wald, Alan. 1987. Theorizing Cultural Difference: a Critique of the 'Ethnicity School'. *MELUS* 14 (2): 21–33.

Wald, Sarah D. 2016. *The Nature of California: Race, Citizenship, and Farming since the Dust Bowl*. Seattle: University of Washington Press.

White, Gwen W., Michael Murray, and Sara E. Jackson. 2004. *Ecosystem Shock: The Devastating Impacts of Invasive Species on the Great Lakes Food Web*. Reston: National Wildlife Federation. https://www.nwf.org/~/media/PDFs/Wildlife/EcosystemShock.ashx

Wilson, Edward O. 1984. *Biophilia*. Cambridge: Harvard University Press.

CHAPTER 6

Facing the End of Nature: Karen Tei Yamashita and Ruth Ozeki

Toward the end of the twentieth century, a few Asian American authors began to move away from the visible distinctiveness that seemed to both anchor and define Asian American culture. Until the 1990s, Asian American literature had explored the dynamics of immigration, had grappled with racism in its multiple forms—legal exclusion, social ostracism and the emergence of "bachelor" societies, the infringement of basic civil rights in "internment" experience, and so on—and had taken part in resistance movements "claiming America" from a cultural-nationalist perspective. Feminist critics (Cheung 1990; Sau-ling Wong 1993) were the first to launch an academic attack against cultural nationalism when they denounced the exclusionary, masculinist bias of the Asian American movement. A second group of Asian American scholars also writing in the 1990s (Lowe 1991, 1996; Palumbo-Liu 1999) started to focus on heterogeneity and diasporic issues, radically questioning the nationalist project precisely because it was nation-locked and not open to the increasing transnational mobility of Asian/Americans. This "denationalization" impulse, to borrow Sau-ling Wong's term (1995), unsettled and destabilized the prior patterns of immigration, unidirectional and final; instead, it emphasized the new immigrants' ongoing physical or virtual ties with the "homeland," which constitute the basic premise of a diasporic identity (Safran 1991, 83–84). By the beginning of the new millennium the increasing internal heterogeneity of the Asian American "constituency" had become too conspicuous to be ignored. The very "ethnoracial"

© The Author(s) 2020
B. Simal-González, *Ecocriticism and Asian American Literature*,
Literatures, Cultures, and the Environment,
https://doi.org/10.1007/978-3-030-35618-7_6

makeup of the Asian American communities had been changing for decades, with the influx of refugees from countries like Vietnam, Cambodia, or Laos in the 1970s and 1980s; the arrival of "global citizens," often highly qualified professionals from South Asia and Southeast Asia (India, Hong Kong, Singapore, etc.), especially since the 1990s; or the rising number of "transethnic/transracial" families, including "mixed-race" individuals and transnational adoptees from Asia. Such protean diversity made people wonder what exactly constituted Asian America in the twenty-first century.[1]

By the 1990s, the increasing sociocultural heterogeneity outlined above was beginning to seep into Asian American literature, as more and more authors chose to publish literary works that apparently departed from the usual Asian American cultural script and focused instead on transethnic and global matters. *Through the Arc of the Rain Forest,* published in 1990 by Japanese American writer Karen Tei Yamashita, was probably one of the first books to openly disrupt conventional understandings of what constituted Asian American literature. In fact, the novel differed so much from earlier Japanese American models—often dealing with the "internment" experience or, more generally, with the racism encountered in the United States—that it was not initially recognized as Asian American. As King-kok Cheung perceptively noted, authors like Yamashita were prone to be "elided in both ethnic studies and multicultural studies" because of the apparent lack of so-called ethnic content in her books (1997, 19).[2] Contemporary reviews confirm that disorientation and surprise, vaguely tinged with unease, pervaded the first reactions to texts like Yamashita's novel. In the last decades we have witnessed a proliferation of other "unclassifiable" books penned by Asian American authors. Writers like Chang-rae Lee, Sigrid Nunez, Jhumpa Lahiri, Monique Truong, or Ruth Ozeki have been moving further and further away from the Asian American script, by choosing to focus on non-Asian American characters and by incorporating a plethora of topics in their fiction, among them environmental and global concerns. This is the reason behind my decision to devote this last chapter to novels like Yamashita's *Through the Arc of the Rain Forest* (1990) and Ozeki's *A Tale for the Time Being* (2013), whose transethnic and (eco)global scope dislocates and problematizes the very "essence" of Asian American literature.[3]

Just as the new century posed serious challenges to the field of Asian American studies, leading some scholars to wonder about the possible demise of Asian America as a concept, by the late 1990s the field of environmental criticism, as outlined in Chap. 1, had also started to face tensions deriving from the slippery nature of "nature" itself. Since the late twentieth century there have been announcements of the impending end of nature. Fredric Jameson, in his *Postmodernism, or, the Cultural Logic of Late Capitalism* (1984), was one of the first intellectuals to establish the present time as the "historical moment of a radical eclipse of Nature itself" (70–71). In his influential *The End of Nature* (1989), Bill McKibben would elaborate on the idea of the death of nature and would announce the advent of a new era, what would later be described as the Anthropocene (Crutzen and Stoermer 2000),[4] the age of "transnatural" nature (Simal 2010), or the advent of "Post Nature" (Clark 2014). In fact, as Ursula Heise eloquently puts it in "Martian Ecologies and the Future of Nature" (2011), "the trope of the 'end of nature' has become one of the hallmarks of the transition from modernity to postmodernity, and of the emergence of a global human presence that leaves no particle of nature unaltered" (451).[5]

Indeed, we may actually be facing *both* the end of "nature" as a transparent concept *and* the end of "physical nature" understood as a more or less untouched nonhuman realm. This may be less serendipitous than it seems, since it was the growing awareness of environmental degradation that forced us to rethink the very nature of nature and our relationship with it.[6] With their respective objects of study questioned and problematized, both Asian American studies and environmental criticism have had to reinvent themselves in order to survive. Novels like Yamashita's and Ozeki's are instrumental in helping to open these new paths. Just as Yamashita's and Ozeki's narratives point at the challenges that Asian American literature will have to face in this new millennium, these novels also epitomize the different challenges that ecocritical theory is currently facing.

1   ECOCRITICISM AND THE END OF NATURE[7]

Much has been written about the idea of nature, from the early treatises by Greek philosophers to Timothy Morton's *Ecology Without Nature* (2007). "Nature" has traditionally conjured up images of more or less "unspoiled" nonhuman contexts such as woods, jungles, mountains, and so on. At the same time, we human beings are often viewed as part of "nature" as well. Not only that, but nature, as Bruno Latour claims, has always been nature-culture (1993, 7). Try as we may, we cannot get away from this aporetic situation, which underscores the radical indeterminacy of the very notion underlying ecocritical analysis. From the beginning, therefore, "nature" has proved problematic, for the term itself is a "fictive" or "discursive" construct (Buell 2003, 43; Mazel 2000, 187), "a notorious semantic and metaphysical trap" (Marx 2002, 30); in sum, "nature" is a nominal nightmare, hardly possible to define, fully laden with "human history," "complicated and changing" (Williams 1980, 284–286).[8] Too narrow an understanding of nature and the role of ecocriticism, as different critics have warned (Phillips 1999, 586; Bartosch 2016, 263), can result in a didactic reductionism that would significantly diminish the imaginative and critical potentialities of literature. Too vague an understanding of the object of study as the (more than natural) environment can likewise prove problematic. Despite the pitfalls that the field has to confront today, there is ample consensus among literary critics and theorists that ecocriticism is here to stay. However, there is also the shared belief that even though much has been attained in the last decades as regards both the popularization and the academic prestige of environmental criticism, the ecocritical movement is bound to face several important obstacles.

In the introduction (Chap. 1) I provided some graphic models to help us understand the trends in ecocriticism in the first decade of the twenty-first century, trends that also allowed us to recognize the challenges of environmental criticism in the near future. The main problems facing ecocriticism, according to Greg Garrard (2004, 2011), are "the difficulty of developing constructive relations between the green humanities and the environmental sciences" and "the relationship between globalization and ecocriticism," which had not been sufficiently addressed by ecocritics in the twentieth century (2004, 178). Therefore, at least two major challenges seemed to accost ecocriticism, the transdisciplinary and the transnational ones; to these two I would add an even more important challenge, that of dealing with a new, "transnatural" understanding of nature.

The Transdisciplinary Challenge

The very origin of ecocriticism, as Glen Love explains in "Ecocriticism and Science" (1999), lies in the need to establish interdisciplinary bridges or, echoing Meeker's words, "the growing need among people … to find a sense of integrity for their own lives and for their understanding of the world around them" (561). Nevertheless, until quite recently, very little real interdisciplinary work has been done that affects the "green sciences" and the "green humanities," mostly due to our mutual distrust and imaginative and intellectual shortcomings. Phillips is rather critical of the field's purported interdisciplinary nature, because he considers that "most of ecocriticism's efforts at being interdisciplinary have been limited to troping on a vocabulary borrowed from ecology" (2003, ix), whereas true interdisciplinarity, according to Roland Barthes, "cannot be accomplished by the simple confrontation of specialist branches of knowledge," it only becomes effective "when the solidarity of the old disciplines breaks down" (qtd. in Phillips 2003, ix).

Despite recent efforts of associations like Association for the Study of Literature and the Environment (ASLE) and European Association for the Study of Literature, Culture and the Environment (EASLCE), there continues to be a need not only to complement knowledge and insights gained from both science and the humanities but to move *beyond* the traditional division of disciplines itself. In fact, my lexical choice, *transdisciplinary*, hints at something qualitatively different from interdisciplinarity. In *Theory, Pedagogy, Politics: The Crisis of the "Subject" in the Humanities* (1991), Mas'ud Zavarzadeh and Donald Morton contend that it is not enough to resort to different fields of knowledge in an interdisciplinary effort, for that would merely engage in "liberal, pluralistic negotiations among knowledges," essentially different from transdisciplinarity, understood as "the locus of politics of knowing and the site of interrogation of the power/knowledge relations of culture" (10). From a different perspective, the theoretical physicist Basarab Nicolescu (2007) also explores the origins and uses of transdisciplinarity. Nicolescu concludes that while both multidisciplinarity ("studying a research topic in not just one discipline only, but in several at the same time") and interdisciplinarity (involving "the transfer of methods from one discipline to another") continue to work "within the framework of disciplinary research," a genuine transdisciplinary approach "concerns that which is at once *between* the disciplines, *across* the different disciplines, and *beyond* all discipline" (2007, 38–9). Therefore, whereas interdisciplinary and multidisciplinary approaches are necessary in ecocriticism, the ultimate goal would be to achieve transdisciplinariness.

The Transnational Challenge
In "Ecocriticism and the Transnational Turn in American Studies" and in *Sense of Place and Sense of Planet*, both published in 2008, Ursula K. Heise stresses the relevant role of localism in the emergence of environmental criticism and the concomitant suspicion of and resistance to—at least on

the part of first-wave ecocritics—anything vaguely reminiscent of globality, even when the incipient social environmentalism of the 1960s and
1970s was paradoxically linked with "global" slogans and icons (2008a,
383–7; 2008b, 28–49; see also Heise 2013). Although it took some time
for the transnational turn to take hold, by the time Garrard pointed out
the challenge of globalization (2004), the global trend was in full swing.
Indeed, in the almost contemporary *The Future of Environmental Criticism*
(2005), Buell expressed the conviction that this global shift was already
taking place and that it would acquire more and more visibility and prominence in the coming years. On a more pessimistic note, Heise continued
to bemoan the absence of a radical reworking of previous paradigms and
the lack of theoretical depth among ecocritics who purported to engage
with twenty-first-century challenges, including the sweeping changes
brought about by globalization.[9] For her, things had not changed fundamentally, since "the theories of subjecthood and agency" underpinning
ecocritical work still failed to "incorporate the changes" triggered by contemporary globalization, "whereas they [did] extensively and with great
philosophical sophistication reflect on the modes of inhabiting local environments" (2008a, 386). To those ecocritics that only superficially endorse
the transnational turn, we have to add those who openly reject it or remain
skeptical of the global approach,[10] despite the fact that the very concept of
global interdependence is implicit in the basic environmentalist tenet of
ecological interdependence.

The Transnatural Challenge

The geographic, literal broadening of ecocriticism, signified by the global
shift outlined above, has been accompanied by a conceptual broadening and
problematization of the very notion of the environment; therefore, it makes
more sense to add this "trans-natural challenge" (Simal 2010) to the transdisciplinary and transnational challenges originally outlined by Garrard
(2004). Over the last two decades, environmental criticism, while not aban-
doning the study of traditional texts like nature writing (e.g. accounts of the
wilderness and "unspoiled nature"), has turned its attention to urban settings (e.g. Gamber 2012), sometimes plying a comparative lens so as to
analyze rural and "wild" natural environments together with less "pristine"
ones. Even the nature of "nature" has been thoroughly reexamined and
questioned, most notably the slippery boundaries between what is deemed
"natural" and what is considered "human-made." What the term "transnatural" foregrounds is the fact that "natural elements" are continually

interbreeding with categories other than "natural." In such a context, the choice of the term "transnatural" keeps reminding us of the fluid and constructed character of the nature-culture, natural-artificial divide.

As might be expected, the excessive attention to the cultural and linguistic mediation in recent environmental criticism has also fuelled the reverse sentiment. Ecocritics like Kate Rigby insist on the continuing presence of "nature" understood as an extradiscursive reality: "the fact that ever more of the earth's surface is currently being refashioned by *techne* does not mean that *physis* has ceased to exist" (2015, 126). However much nature may have been altered or transmogrified, nature is far from dead; and the same can be said of ecocriticism. Still, after the "linguistic turn" in theory, with its poststructuralist assault on the positivistic approach to extralinguistic "reality," it has become more and more difficult to sustain a simplistic understanding of "nature."

In reaction to the poststructuralist deconstruction, or at least partial questioning, of nature itself, some ecocritics have started to focus once more on the very materiality of our bodies and our environment, a theoretical shift that has been dubbed the material turn (Iovino and Oppermann 2012; Iovino and Oppermann 2014a).[11] Serenella Iovino and Serpil Oppermann define "material ecocriticism" as a new critical paradigm born of the confluence of the new philosophical materialism and postmodern ecological theories (2012, 448). The objective of material ecocriticism would be to construct "new conceptual models" that allow us "to theorize the connections between matter and agency" and "the intertwining of bodies, natures, and meanings" (2012, 450). Material ecocritics would thus "read" matter as text, for their basic assumption is that the material world is graspable as a narrative with its own logic: "the world's material phenomena are knots in a vast network of agencies, which can be 'read' and interpreted as forming narratives, stories" (Iovino and Oppermann 2014b, 1).[12]

This renewed attention to matter and to our own material limits has also blossomed in the form of genres such as "toxic discourse" (Buell 1998) and "global warming" literature.[13] The rising popularity of "toxic" narratives and environmental dystopias like the cli-fi subgenre in the last decades corroborates that toxic discourse has become as pervasive as pollution and toxic waste. Lawrence Buell's thesis is that the increasing corpus of narratives about toxicity shares a distinctive rhetoric, "an interlocked set of topoi whose force derives partly from the exigencies of an anxiously industrializing culture, partly from deeper-rooted Western attitudes" (1998, 639).[14]

In fact, in some of the most influential novels written in the 1980s, Cynthia Deitering saw a growing awareness that the "pristine" character of nature was no longer there. In "The Postnatural Novel" (1992), Deitering argued that, in the course of the decade, there had been a significant "shift from a culture defined by its production to a culture defined by its waste" (196). Accordingly, the fiction written in those years showed signs that Americans were becoming aware not only of the toxic waste surrounding them but also of their own complicity in the degradation of the planet and the proliferation of "riskscapes" (200).[15] These riskscapes are by no means a thing of the past, but continue to haunt us in the twenty-first century. If Deitering focuses on the growing worry about chemical toxicity that filled the novels published in the 1980s, Molly Wallace's recent book, *Risk Criticism*, discusses how pollution and toxic waste are still a very pressing concern in recent cultural production. Among other narratives, Wallace reads Ozeki's *All Over Creation* as "an example of risk realism, a novel that grapples with the question of what it might mean to dwell in our present condition of uncertainty" (2016, 98). John Gamber offers a more sanguine vision of toxicity in *Positive Pollutions and Cultural Toxins*, where he analyzes several narratives published by "ethnic" US writers in the last decades. One of the texts that the critic chooses as a paradigmatic example of "positive pollutions" is Yamashita's *Tropic of Orange*, inasmuch as the novel shows boundaries as "porous, and constantly permeated and penetrated," thus illustrating the need to go beyond or at least question "absolutes of purity" (2012, 122).[16]

Both Ozeki's and Yamashita's work provide fertile ground for an exploration of the recent trends of ecocritical theory outlined above, including the emergence of toxic and risk criticism. In the first section of this chapter I will show how Yamashita's *Through the Arc of the Rain Forest* revisits and modifies previous discourses, most notably the pastoral mode, in an attempt to meet the new transnational and transnatural challenges outlined above. In this novel, the devastating dynamics of the "waste of the empire" (Deitering 1992, 199), implicit in the allusion to the polluting transnational corporation (Georgia and Geoff Gamble, or GGG), are fleshed out in the trope of the Matacão. Similarly, Ozeki's novels openly grapple with the material consequences of the capitalist model exported by America to the rest of the world, as we shall see in the second section of the chapter. Whereas Ozeki's environmental agenda is rather obvious in her first two books, which indict the meat industry and GMO corporations, respectively, in the context of global capitalism, in her third novel her focus moves away from toxic agribusiness practices. Although she continues to explore issues of global interdependence and connectivity, in *A Tale for the Time Being* she seems as interested in metaphysical concerns as in material ones. In the last part of this chapter, I will try to demonstrate how Ozeki's particular handling of different time-space paradigms in this novel allows her to reconcile materiality with immateriality.

2   KAREN TEI YAMASHITA'S *THROUGH THE ARC OF THE RAIN FOREST*: TRANSNATIONAL-TRANSNATURAL POST-PASTORALISM[17]

where are we to classify the ozone hole story, or global warming, or deforestation? Where are we to put these hybrids? Are they human? Human because they are our work. Are they natural? Natural because they are not our doing. Are they local or global? Both. (Bruno Latour, *We Have Never Been Modern*, 50)

Maybe the 'trans' is what my writing is. (Yamashita, interviewed by Te-Hsing Shan 2006)

Yamashita's *Through the Arc of the Rain Forest* not only illustrates the interpenetration of the local and the global, the "transnational challenge" that I have previously outlined, but it also broaches the transnatural challenge, with abundant examples of the reworking of the culture-nature divide. Foremost among these revealing episodes and motifs are the novel's double trope of the ball-Matacão, on the one hand, and the discovery of a junkyard in the jungle, on the other.

2.1   *Plastic Flesh*

Through the Arc of the Rain Forest constitutes an apt illustration of the recent developments in environmental studies, which have broadened what we mean by nature and the environment,[18] at the same time that they rehearse the movement from the local to the global. Yamashita's novel not only includes a US-based transnational corporation (GGG), an obvious reminder of contemporary economic globalization and its effects on the environment, but also features an international cast of characters. Thus, in *Through the Arc of the Rain Forest* we find an Amazonian farmer who becomes an expert in "featherology" (Mané), a young boy from the North of Brazil (Chico Paco) who specializes in pilgrimages and becomes some sort of evangelical radio preacher, an urban Brazilian couple (Batista and Tania Aparecida) who start breeding pigeons for a living and end up building a pigeon-communication empire, an American businessman with three arms (J.B. Tweep), a French scientist with a "trio" of breasts (Michelle), and Kazumasa Ishimaru, a Japanese man who has recently migrated to Brazil and has brought with him a rotating miniature planet, a ball that constantly whirls just inches away from his forehead. Surprisingly, it is this ball, invariably attached to Kazumasa since childhood, that assumes the narrative voice and tells us the convoluted story of how the paths of these multifarious characters crisscross and converge at a single symbolic location, the "natural" wonder of the Matacão plateau, suddenly emerging in the middle of the Brazilian rain forest. Arguably, although the novel is mostly circumscribed to a single national setting, Brazil, the main events occur at a site, the Matacão, that is profoundly global(ized).

The unfathomable Matacão expanse is made of an unknown, highly resilient material, the same kind of plastic which, we later learn, also constitutes the "flesh" of Kazumasa's personal satellite.[19] The Matacão is "neither rock nor desert [...] but an enormous impenetrable field of some unknown solid substance stretching for millions of acres in all directions" (Yamashita 1990, 16). The Matacão eventually turns out to be the eruption of non-biodegradable waste from industrialized countries in the middle of the rainforest. Yamashita's inclusion of this anthropogenic formation is a trenchant criticism of what Deitering deftly calls "the waste of the empire" (1992, 199). In order to clean their own country, as Buell described in "Toxic Discourse," Americans exported both polluting industries and toxic waste, which were relocated as "maquiladoras along the U.S.-Mexican border, sweatshops in Latin America and southeast Asia, garbage flotillas to Africa," thus increasing "eco-inequality on a global scale" (Buell 1998, 644).[20]

The Matacão not only constitutes an indictment of ecological imperialism, it also introduces the concept of transnatural nature. In fact, it could be argued that the emergence of the Matacão clearly turns the wilderness of the Amazonian rainforest into a hybrid bionetwork, an overlapping of organic and inorganic realms, an *anthrome*.[21] Due to its versatility and resilience, the "miracle plastic" is greedily mined and extracted from the Matacão site, and all sorts of products are made out of the new, apparently indestructible material, inaugurating the Plastics Age (Yamashita 1990, 143). Like the Matacão itself, many of the plastic artifacts made of Matacão material seem about to traverse the boundary that separates the natural from the artificial, even the living from the nonliving. Plastic, for instance, imitates feathers (157), body parts (142–143), food (143), and all sorts of plants and animals like palm trees, zebras, or lions (168). The Matacão is the perfect simulacrum indeed, as seen in the "plastic zoo" included in Chicolándia: "The animated animals, also constructed in the revolutionary plastic, were mistaken for real animals until people questioned their repetitive movements, their obviously benign nature, and the trade-off in smells: the warm stench of animal refuse for a sort of gassy vinyl scent" (168).[22]

The Matacão plastic copies in Yamashita's novel become even more convincing when trying to reproduce more static beings such as rooted plants. They come even closer to perfection than "the real thing" and, paradoxically, they prove more apt at conveying "the very sensation of life" than their living counterparts:

> At the plastic convention, two tiger lilies, one natural and the other made from Matacão plastic, were exhibited for public examination. Few, if any, of the examiners could tell the *difference between the real and the fake*. Only toward the end of the convention, when the natural tiger lily began to wilt with age, bruised from mishandling, were people able to *discern reality from fabrication*. The plastic lily remained the very perfection of nature itself. Matacão plastic managed to recreate the natural glow, moisture, freshness— the very sensation of life. (142; emphasis added)

The flaw of the argument, of course, lies in the fact that nature is not necessarily perfect, in the sense that it involves physical deterioration, violence, and death. Indeed, the very perfection of nature is not "natural." Such a vision of nature is, as all constructs, human-made.

Together with the mystifications linked to the perfection of nature and the momentary confusion of real/plastic life, the very origin of the Matacão contributes to the erosion not only of the boundaries between the natural and the artificial but also of those between the local and the global, as several critics have pointed out. In "'A Bizarre Ecology': The Nature of Denatured Nature" (2000), Molly Wallace argues for an understanding of postmodern ecology that does away with the dualistic vision of a nature-culture dichotomy. For Wallace, Yamashita's novel builds bridges across traditional divides, by proffering "an ecosystemic vision of nature and culture which provides a model for and a critique of hybridity" (2000, 149).[23] In "Local Rock and Global Plastic," Ursula Heise addresses the manner in which the phenomena of "disembedding" (Giddens) and deterritorialization (Tomlinson) figure in *Through the Arc of the Rain Forest*, as well as the links "between ecological and cultural globalism" (2004, 127) that Yamashita forges in her novel. The fact that the Matacão plateau turns out to be the result of the combination of human interaction with the biotic community all around the world, even though its emergence is restricted to (apparently) a single locale, constitutes "a striking trope for the kind of deterritorialization John Tomlinson analyzes, the penetration of the local by the global that leads to the loosening of ties between culture and geography" (135). However, while Yamashita successfully addresses "the ambiguities of an ecologically based sense of place in the age of globalization" (132), for Heise, the Asian American author eventually circumvents the need to provide adequate answers to the ecological problems raised in the book by proffering a sociocultural solution instead (139; cf. Gamber 2012, 12–13).

Nature and human actions and global and local threads crisscross and get entangled in the Matacão. Well into the novel we learn that the Matacão expanse is the result of huge "landfills of non-biodegradable material buried under virtually every populated part of the Earth" which, under "tremendous pressure," had "pushed ever farther into the lower layers of the Earth's mantle," from where these "liquid deposits of the molten mass had been squeezed through underground veins to virgin areas of the Earth" (202). If the transnational theme was already present in the myriad of connections between people (Kazumasa, Mané, Tweep, Chico Paco, etc.) and human institutions (GGG) within and across nations, the planetary dimensions of a localized phenomenon, both in its origins and in its consequences, only emphasize the global paradigm even more. Similarly, the very formation of the Matacão plateau necessitates both human activities—mostly the excess they generate in the form of non-biodegradable waste—and natural geological forces, in this case the pressures and movements that take place under the external layer of the planet. Both combine to create a material that, although anthropogenic, cannot be totally disengaged from nature.[24]

At this stage, then, we cannot but wonder whether the traditional natural-artificial dichotomy still remains in place in our postmodern, transnatural world. With the advent of poststructuralism, it has become hardly possible for any critical school to embrace the essentialist naivety that is so noticeable in early ecocriticism. Can we actually locate some element or thing that has not been tampered with or directly produced by human beings and human discourse? Leo Marx indirectly points at this dilemma in his rhetorical question about the apparent paradox and clash between the machine and the garden: "If technology is the creation of man, who is a product of nature, then how can the machine in the landscape be thought to represent an unresolvable conflict?" (1964, 242). In "Ideas of Nature," Williams explicitly argues against trying to sever "nature" from human action, since, then, "it even ceases to be nature" (1980, 294). In a similarly paradoxical phrasing, Williams explains how the more human-nature interactions increase, the more necessary the separation between both entities becomes (1980, 295–296). In the end, it does not make much sense to insist on the dichotomy people versus "nature" or "natural" versus "artificial": human beings "have mixed our labour with the earth, our forces with its forces too deeply to be able to draw back and separate either out" (1980, 296).[25] It is by carefully looking into the specific material practices that complicate the interactions between humans and "nature" that we may arrive at some honest reassessment of the situation (1980, 296–297).

These proto-ecocritical explorations of the inextricability of nature and culture reappear in later theorists, like Bruno Latour, who claim that in the contemporary world, nature and culture are so imbricated with each other that it is hardly possible to tell one from the other. He provides telling examples of this in *We Have Never Been Modern* (1991/1993), where he asks readers: "where are we to classify the ozone hole story, or global warming, or deforestation? Where are we to put these hybrids? Are they human? Human because they are our work. Are they natural? Natural because they are not our doing" (50). These phenomena are both "human" and "natural," or, using my terminology, they are "transnatural" in that they problematize the artificial/natural dichotomy. Texts like Yamashita's novel force ecocritics to engage in a revision of the nature-culture divide by approaching both poles of the dichotomy "as interwoven rather than as separate sides of a dualistic construct" (Armbruster and Wallace 2001, 4). Yamashita's central trope of the Matacão, the result of both human intervention and geological forces, transmogrifies nature and renders it "transnatural." Not only that, but the discovery of a "transnatural" plateau in the middle of the Amazonian rainforest, the epitome of "pristine nature," where one would expect a "natural" landscape, combined with the presence of plastic substitutes for natural feathers, plants, or food further disrupts conventional understandings of what is natural. As Phillips insightfully argues, "encounters with artifice, where one expects to find the real thing," point to the fact "that we have found a substitute for 'the natural world': in the postmodern world, nature no longer seems to be necessary" (1994, 215).[26]

The very question of what is natural and what is artificial, as raised by the Matacão plastic, is prefigured in an apparently insignificant episode at the beginning of the novel: the origin of sand bottling. The narrator tells us how the first innocent gesture of filling a bottle with multicolored sands as a nostalgic memento of one's birthplace is soon co-opted and marketed as tourist souvenir:

> A young talented boy had then gotten the idea of pouring the colored sand in bottles in such a way as to create pictures. [...] One day, a tourist brought a picture of the Mona Lisa and asked the boy to duplicate it in a sand bottle, and he did. After that, the boy left the town and went away to be famous, sand-bottling every sort of picture from the President of the Republic to the great Pelé. Someone said he no longer used real sand but some synthetic stuff dyed in every color you could imagine. Someone said he was even making sand pictures in bottles of fine crystal and mixing the sand with gold and silver dust. (24–25)

The drift from the simple to the elaborate and from the natural to the artificial is here described as the shift from the "real" to the "synthetic," obliquely presented as non-real, that is, ambiguously construed as either (postmodern) virtual or (premodern) magical.[27]

The same contradictory connotations surround the novel's main leit-motifs: the feathers and the Matacão. Feathers present at least two para-doxes: they are both from a living creature, but not living themselves; they are as ancient as birds, but also the latest rage/fad in contemporary soci-ety; they seem to restore human health and harmony, but "deceitfully" so (185), since they finally bring about much suffering and death. Likewise, the central natural-artificial trope, the Matacão, is both adored as some ancient substratum erupting from the depth of the earth and praised as the ultimate, modern scientific discovery. In fact, as mentioned earlier, Matacão plastic becomes the postmodern material par excellence, a superbly malleable simulacrum, a "virginal" product which the transna-tional corporation GGG soon eyes with equal amounts of perplexity and greed, until it literally becomes "plastic money" (141):

> The wonderful thing about the Matacão plastic was its capability to assume a wide range of forms. ... Every industry from construction to fashion would jump into Matacão plastics. At a plastics convention, all sorts of marvels were displayed—cars made completely out of Matacão plastic, from the motor to the plush velveteen fabrics of the seats; imitation furs and leathers made into coats and dress pumps; Danish furniture made of Matacão "teak"; and all sorts of plants, from potted petunias to palm trees. The remarkable thing about Matacão plastic was its incredible ability to imitate anything. (142)

In spite of such malleability, this apparently non-biodegradable plastic that excels at imitation—like the junkyard in the jungle, as we shall shortly see—is eventually swallowed by and integrated into "nature," enacting the symbolic—though not uncomplicated—"return" that the title of the final section hints at. The Matacão plastic, which seemed to be immune to all life forms, eventually falls prey to some mysterious bacteria, which literally gnaw it away. The whole range of products, from clothes to food, made of Matacão plastic soon start deteriorating and finally crumble down, and with them the plastic empire itself. The first sign of such destruction can be seen in the narrating ball and its "plastic flesh." One day Lourdes and Kazumasa notice that the swirling ball looks more "lopsided" and less spherical than usual; indeed, the ball claims:

> I seemed fraught with tiny holes. .... Something was eating me, carving out delicate pinhole passages, which wound intricately throughout my sphere. ... Every day, Kazumasa watched more and more of me disappear, my spin grow slower and more erratic. ... One day, he touched me tenderly and was shocked to find his finger pierce the now very thin veneer of my surface. Within, I had been completely hollowed out by something, by some invisible, voracious and now-gorged thing.
> The next day, Kazumasa awoke and wept uncontrollably at the unobstructed view of the room before him. (205–206)

The oxymoron of "plastic flesh" intuited both in the person-cum-ball known as Kazumasa and in the very ambiguity of the Matacão has already been encoded in cryptic omens ensconced in every nook and cranny of the novel: in the reference to the "plastic snakes" which vendors sell on the street (35); in the GGG memos about "Natural vs. Plastic Plants and Employee Morale" (52) or about the viability of "artificial non-polluting snow" (54); in the contrast of the "dense concrete jungle" with the Amazonian "living jungle" (82); in the depiction of GGG as "a great functioning miracle, a living, breathing organism" (111); or in the very paradox of "nature" imitating human artifacts in the junkyard ecosystem (100). The oxymoron natural-artificial reappears in a more explicit way when we witness how the "plastic paradise" of Chicolándia, including its plastic plants and animals, becomes "horribly disfigured, shot full of tiny ominous holes, the *mechanical entrails* of everything exposed beneath the once-healthy *plastic flesh*" (206–207; emphasis added). If the artificial and the natural had become momentarily confused, now the paradoxical combinations of "mechanical entrails" and "plastic flesh" cannot leave anyone indifferent. By juxtaposing lexical items that apparently belong to opposing semantic fields, these phrases call our attention to the difficulty of drawing a neat boundary between the animate and the inanimate, the "natural" and the "(hu)man-made." However, it is in the description of the aforementioned junkyard ecosystem that the paradox of "mechanical entrails" and "plastic flesh" is articulated in all its complexity; it is there where the emergence of an "artificial nature" becomes especially notorious.

2.2   *The Junkyard in the Jungle*

Just as the sand bottle episode prefigured the oxymoronic plastic flesh, the huge dump that the Matacão turns out to be finds a smaller replica in the "natural-artificial" ecosystem born around an unusual metal cemetery, where abandoned planes and cars coexist with sentient beings. Yamashita's "ecological experiment" has been interpreted as an apt commentary on the nature-culture divide, thus reinforcing the issues previously raised by the Matacão (Heise 2004, 144–145). Heise deftly explains how, in the description of this peculiar ecosystem, Yamashita has exaggerated what are, in principle, "basically plausible processes of adaptation," and she has done so in order to emphasize both "the transformability of biological species and their partly natural, partly technological environments" (145). For Jinqi Ling, the metal cemetery episode most prominently functions as a painful reminder of "systematic imperialist violence" on the part of the United States, since the abandoned war aircraft and other vehicles, all US-made, bring echoes of "the post-WWII U.S. hegemonic control over Latin America" (2006, 10; cf. Wess 2005, 110; Bahng 2008). Complementing these various interpretations, I will argue that the particular bionetwork created around the metal cemetery can be read as "the junkyard in the jungle" in the light of Leo Marx's *The Machine in the Garden*.

Marx's influential book mulls over the multiple ways in which American literature records the irruption of technology in what had long been regarded as a natural, even virginal, landscape. The author first explores the tropes of the garden and the machine in a separate fashion before putting them side by side as "two kingdoms of force," echoing *The Education of Henry Adams*. Marx maintains that the pastoral tradition, with its emphasis on its escape from urban life and its retreat into "nature," acquires a special import in American literature, as America has traditionally (and, we would add, Eurocentrically) been construed as an Edenic new land, the Brave New World indeed. Marx particularly focuses on what he terms the "Sleepy Hollow" moment, where the intrusion of a machine (or its surrogate) interrupts, and disrupts, a pastoral moment (1964, 11–16).[28] In order to render this literary topos less nationally specific and more transnational, I have rechristened Marx's "interrupted idyll" with a name deriving from another famous Latin literary topos: *locus amoenus truncatus*.

Marx's thesis is both echoed and turned on its head in Yamashita's description of "the junkyard in the jungle." Halfway through the novel we learn from the narrator that a huge "parking lot" has been discovered in the middle of the Amazonian rainforest, a space full of abandoned planes and cars, wrapped up in "criss-crossing lianas [that] completely engulfed everything" (99). One part of this peculiar parking lot now contains "a large pit of grey, sticky goop," composed primarily of napalm (99). Incredible though it may seem, this idiosyncratic anthrome, the anthropogenic ecosystem of the jungle junkyard, has actually produced new species of fauna and flora, among them mice that "burrowed in the exhaust pipes," wrapped up in "splotchy greenandbrown" hair (mimetically camouflaging themselves in old military vehicles) or else in "shiny coats of chartreuse, silver and taxi yellow" (imitating the colors of other cars and planes); a "rare butterfly" whose "exquisite reddish coloring" is given by "a steady diet of hydrated ferric oxide, or rusty water"; the new air plant with "carnivorous flowers" that "attached itself to the decaying vehicles"; or the tribe of monkeys that "had established territory in the carcasses of the bomber planes" and has shot another tribe to extinction (100–101).[29]

Should we apply Marx's grid to this significant scene, we would record not only the obvious similarities but also the ways in which this episode departs from the master theory. In this case, the abandoned parking lot is found by teams of entomologists who, in a neo-pastoral move, try to escape from their urban environments in search of "genuine" nature— more specifically, in search of a rare butterfly. The physical intrusion or "emergence" of a junkyard full of old vehicles in the middle of the apparently pristine, virgin rainforest not only interferes with their pleasure but totally disrupts the pastoral moment: an obvious case of *locus amoenus truncatus*. Nevertheless, despite the ostensible similarities, one cannot ignore the differences between Marx's "interrupted idylls" and Yamashita's peculiar brand of "the machine in the garden." To start with, Yamashita's chosen location is a jungle, not a "garden," a middle landscape in between "civilization" and "wilderness." In addition, we cannot fail to notice that the setting for this episode, and for most of the novel, is not the United States, but Brazil. Last but not least, in *Through the Arc of the Rain Forest* it is not a noisy, active machine that interrupts the immersion in nature, but a heap of old, useless cars and planes, silent and passive.

It could be argued that the first departure from Marx's theoretical model can be easily bypassed. Granted that, more often than not, the *locus amoenus* of the pastoral mode used to be the garden or some other version of the middle landscape, not a "virginal" rainforest. However, a careful reading of the Amazonian forest as portrayed in Yamashita's book leads us to the realization that, after the discovery of the Matacão, the rainforest has been "tamed" to such an extent that it comes closer to a domesticated middle landscape, a garden, than to "pure" wilderness. By the same token, the next divergence could be circumvented by noting that Brazil has equally been construed as a virgin land, the New World, the occasion for a rebirth of countless immigrants, among them Kazumasa. However, while America as a continent remains the setting for *Through the Arc of the Rain Forest*, it is clearly not the America that Marx had in mind, the United States of America.

The change in location involved in this second departure poses new, interesting challenges, most notably because Yamashita's work is usually labeled Asian American. As mentioned at the beginning of this chapter, after decades of "claiming America," the field of Asian American studies has undergone a gradual process of "denationalization," most visible since the mid-1990s. And this is not an isolated phenomenon, one that only affects certain minorities. The deterritorialization attendant in recent globalizing processes has "unmoored" (Lee 2007, 503) all types of national and ethnic subjecthood. The advent of contemporary globalization has rendered traditional national/ethnic identities if not obsolete, at least in need of revision (Koshy 2005, 111, 117).[30] It cannot be denied that previous understandings of American national identity in early ecocritical work, such as Marx's *The Machine in the Garden*, have to be revised in our globalized world, a task that Yamashita's *Through the Arc of the Rain Forest* and subsequent novels entice readers to do.

Finally, the third departure from the master theory involves the nature of the machines in the junkyard, which will directly determine the ecosystem they are part of. The "living," noisy machines, most notably the train, of the *loci amoeni truncati* described in Marx's *The Machine in the Garden* are here replaced by "dead" machines, a stark reminder of the by-products of consumerist capitalism. In this postmodern, postindustrial context, the junkyard in the jungle becomes a "cyborg ecosystem," where machines and living organisms meet. According to Donna Haraway's famous definition of the cyborg, the latter is "a cybernetic organism, a hybrid of machine and organism," "a condensed image of both imagination and material reality," which renders the frontiers between living organisms (humans, animals, plants) and machines particularly "leaky" (1991, 315, 317). As boundary-trespassing hybrids, cyborgs, including cyborg ecosystems like Yamashita's junkyard, contribute to the problematization of entrenched hierarchies and watertight categories, such as the natural and the artificial. In order to survive in the surprising "cyborg-network" imagined by Yamashita, the new species of mice living in the junkyard adapt to and copy human-made vehicles (military jeeps and planes), which in turn try to imitate nature in their camouflaging green and brown colors (Yamashita 1990, 100). Nature and culture chase each other's tails. Read in the light of cyborg theory, the ecosystem created around the metal cemetery constitutes a further exploration of the dualism of nature versus culture, nature versus technology, confirming the previous insight that such a divide should be understood as interconnected constructs rather than as clear-cut dichotomies.

What transpires after a detailed analysis of both the Matacão phenomenon and the cyborg parallel is the stark realization that we are indeed living in a transnatural world where nothing remains untouched, everything has been directly or indirectly "contaminated" by human actions, and culture and technology have invaded what used to be the inviolable realm of "nature."[31] But the reverse is also true and becomes conspicuous not only in the catastrophic end of the Matacão plastic empire but also in a complementary reading of the trope of the junkyard in the jungle. Once more deviating from the master theory, in Yamashita's novel the abandoned, rusty planes and cars appear to be a monument to the past—or rather, a denunciation of the American imperialist past—not an anticipation of the future, as was the case in *The Machine in the Garden*. Extricated as they are from the utilitarianism attached to machines, the vehicles in Yamashita's novel become "denaturalized" machines that, paradoxically, become "naturalized," that is, literally invaded by "wild nature." On this occasion, therefore, the pastoral idyll may not be totally interrupted, but it acquires unsettling undertones: as dead machines, as technological corpses, these old vehicles are soon "ingested," processed, incorporated, by nature itself, a "composting" that can be read as enticingly positive despite the initial nightmarish shock. Indeed, the machine, or rather the junkyard, made of old machines, becomes part of the land, of the natural or—more accurately—transnatural environment. In a thinly disguised *mise en abyme*, machines serve as the basis for a microecosystem within the larger ecosystem of the Matacão rainforest. In both cases the separation between the natural and the artificial is severely undermined. The "junkyard in the jungle" starts with Marx's intrusion of the machine in the garden, only to have the natural environment adapt to and finally swallow the machine. The garden in the machine.

2.3   Transnational-Transnatural Post-Pastoralism

The preceding analysis of Yamashita's *Through the Arc of the Rain Forest* confirms the widespread intuition that nowadays "fewer and fewer of the world's population live out their lives in locations that are not shaped to a great extent by translocal—ultimately global—forces" (Buell 2005, 63). This "shrinking of the planet" (68) is rendered literal in the minute globe that interpellates us as narrator. Through its "ecospeak" (45), or rather, as befits a globe, its "globalspeak," the ball tells us the story of how the local interpenetrates with the global, and the other way around. From Kazumasa's metaphorical little planet in *Through the Arc of the Rain Forest* to the great shifting of the Tropic of Cancer in the larger planet in *Tropic of Orange* (1997), the effects of globalization have been a recurrent concern in Yamashita's work, which, as Sue-im Lee reminds us, urges readers "to conceive of a new collective subject positioning that can express the accelerated movement of capital and humans traversing the world" (2007, 502).[32] In addition, in *Through the Arc of the Rain Forest*, Yamashita's transnational concerns are further interrogated and compounded with pressing ecological issues which can be read as transnatural, that is, not so much transcending or going beyond nature, as interbreeding it with categories other than "natural." In sum, Yamashita's novel encodes two of the challenges that environmental criticism is currently facing.

At the same time, as I have attempted to demonstrate, a significant episode in *Through the Arc of the Rain Forest* harks back to the paradigm put forward by Leo Marx in the 1960s, "the machine in the garden." At the same time that Yamashita's transmogrified pastoral revisits the old master theory, however, it revamps it not only by destabilizing the classic human-nature divide inherent in first-wave ecocriticism, but also by adding the transnational ingredient. Thus, the machine-in-the-garden paradigm is updated in order to incorporate the broadening of current environmental criticism, both literally (globalization) and conceptually (transnatural nature). While at times Marx's paradigm may metamorphose and undergo a reversal, so that we encounter a peculiar "garden" sprouting from the machine, the old trope also corroborates its continuing validity. Though filtered by the sieve of globalization and shaken by the emergence of cyborg ecosystems, "the machine in the garden" has survived as a compelling ecocritical framework, even if it occasionally mutates into a junkyard in the jungle.

Arguably, Yamashita's novel might be accurately labeled "post-pastoral" fiction. As mentioned in earlier chapters, Gifford (1995, 1999) posited the post-pastoral mode precisely as a contemporary "mutation" of the old pastoral form. As a quick overview will show, Yamashita's *Through the Arc of the Rain Forest* exhibits most of the key features of post-pastoralism. First, contemporary post-pastoral literature fosters an ecological consciousness in the face of growing environmental degradation. Yamashita's "green" agenda is conspicuous, for it sounds the alarm of the very real "postmodern danger of global environmental destruction" (Rody 2000, 629). For Gifford, post-pastoral narratives also acknowledge that "nature" can be as much destructive as creative (1999, 153). In Yamashita's *Through the Arc of the Rain Forest*, the creative aspects of nature can be seen in the healing effects of feathers, among other examples, while the destructive side becomes obvious and palpable in the typhus that annihilates both human (e.g. Mané) and nonhuman animals (e.g. the birds). In addition, post-pastoralism incorporates the insights gained thanks to movements like environmental justice and ecofeminism (Gifford 1999, 165–66). Yamashita's novel hints at the shared oppression of different "Others": "nature," women, ethnoracial minorities, and social subalterns. Another key feature in contemporary post-pastoral fiction is the realization of the inextricability of culture and nature, "an awareness of both nature as culture and of culture as nature" (Gifford 1999, 162). As we have seen in our previous analysis, the very "nature" of the Matacão, as described in *Through the Arc of the Rain Forest*, deteriorates the boundaries between what is natural and what is artificial. When the enigma of its origin is finally resolved, we confirm the intuition that the mysterious substance making up the Matacão is anthropogenic. The bizarre plateau is the upshot of the accumulation of non-biodegradable waste in/from industrialized countries, the Global North, which, like many other negative consequences of contemporary globalization, comes to the light in the Global South.

Last but not least, one of the aims of contemporary post-pastoral literature, Gifford claims (1999, 152–153), is to criticize human *hübris*, much like Leopold or Plumwood had done in their work. At the end of *Through the Arc of the Rain Forest*, as we have seen, the high price paid for human greed is all too visible and humankind is eventually humbled. When the unknown bacteria start eating "the once-healthy plastic flesh" (207), the widespread disintegration that follows is both literal and metaphorical. The "plastic paradise" literally disintegrates and crumbles; the global economic system, embodied here in the stock market, plummets down "as the invisible bacteria gnawed away, leaving everything with a grotesquely denuded, decapitated, even leprous appearance" (206, 207). These bacteria are described as eating the "plastic flesh" and "mechanical entrails" of those Matacão artifacts, a double oxymoron that further develops the theme of the complicated relationship between the animate and the inanimate, the natural and the human-made. However, these phrases can also acquire an ironic import, a narrative strategy employed to denounce the ease with which human beings naturalize what is artificial. In this second interpretation, the resilience of living organisms—in this case plastic-eating bacteria—would therefore become the necessary counterweight that resituates Non-Nature and Nature—the Machine and the Garden, to use Marx's leitmotifs—in their respective separate realms. Yet this traditional conservationist reading is not easily borne out by the ambiguous ending of Yamashita's plastic pastoral, which has invited different and at times contradictory interpretations.

Through the Arc of the Rain Forest ends with several brief vignettes. In the first, Yamashita depicts Chico Paco's long funeral procession across Brazil, "from the Matacao toward the sea" (209). The bleak scenes described as the funeral procession winds its way across Brazil confirm that "the despoliation of nature seems to continue unabated even after the end of the Matacão culture" (Heise 2004, 137). The echoes of a classic of toxic discourse, Rachel Carson's *Silent Spring*, are evident in the description of the procession "as it snaked its solemn way through the now-muted forest" (209).[33] The second and third vignettes are much more hopeful. While the second scene shows Batista's eager anticipation before his reunion with his estranged wife, Tania, in the third vignette we catch a glimpse of the new happy family created by Lourdes and Zazumasa, a scene that approximates the trope of pastoral bliss. In anticlimactic sequence, after this image of Edenic abundance, Yamashita approaches and depicts the present in a less-optimistic, quasi-apocalyptic way. At the same time that the end of the novel intimates the triumph of the natural world, the return of Nature (as in the episode of the junkyard in the jungle), it is no less true that something has been lost in the interval, as explained in the description of what used to be the Matacão plateau:

> But all this happened a long time ago.
>
> Now you may look out across this empty field, strewn with candle wax, black chicken feathers and those *eternally dead* flowers, discarded jugs of can brandy, the dirt pound smooth by hundreds of dancing feet. [...]
>
> On the distant horizon, you can see the crumbling remains of once modern high-rises and office buildings, everything covered in rust and mold, twisted and poisonous lianas winding over sinking balconies, trees arching through windows, a cloud of perpetual rain and mist and evasive color hovering over everything. *The old forest has returned* once again, secreting its digestive juices, slowly breaking everything into edible absorbent components, pursuing the lost perfection of an organism in which digestion and excretion were once one and the same. *But it will never be the same again.* (212; emphasis added)

Heise suggests that the turn of events at the end of the novel argues against reading *Through the Arc of the Rain Forest* "only as an anti-globalization story" (2004, 136). The critic resists the obvious interpretation that the catastrophic events closing the novel "signal the termination of the globalizing project and the return to a more authentic experience of place" (137). Heise insists that the end of the novel goes back to the "pastoral cliché" of "bucolic bliss," offering "a sociocultural solution to a problem [...] earlier articulated in ecological terms" (138). She sums up her argument in one sentence: "Ecological deterritorialization is contained by cultural reterritorialization" (139).[34] My interpretation of this last chapter is rather different, for the image closing the book is not so much that of domestic or pastoral bliss, as the description of the slow recovery of the rainforest, with the crucial caveat that "it will never be the same again." As I have argued elsewhere (Simal 2009, 220–221), instead of resorting to "an Edenic deus-ex-machina resolution," the author has preferred to wield the complex mode of "comic apocalypse."[35] The fact that Yamashita has chosen to place this gloomy depiction of the ongoing threat to the environment—in this case the rich Amazonian forest—at the end of the novel has the immediate effect of renewing the unease among readers. It is that threat that does not let us merely dwell in the apparent happy ending of the transethnic family born at the end of the novel, but instills in us the need to further scrutinize the ways in which we interact with nature on a global scale.

3   RUTH OZEKI'S NOVELS: MATERIALITY AND IMMATERIALITY

Fiction is an elemental force, which has the power to shape reality in its own image—or images, I should say—because reality, like light, exists not only as a single point or particle, but also as an array of possibilities. (Ruth Ozeki, quoted by Flood in *The Guardian* 2014)

3.1   *Meat, Potatoes, and Gyres*

While Yamashita offers a sharp critique of the imperialist dynamics of waste in *Through the Arc of the Rain Forest*, it is probably Ruth Ozeki who most openly tackles the dire consequences that the capitalist model exported by America has for our planet. Indeed, read in conjunction, Ozeki's first two novels can be considered a sweeping indictment of the role played by US-based transnational corporations (TNCs) in the degradation of our planet. As in Yamashita's book, ecological concerns, cross-cultural bonds, and transnational connections are prominent in Ozeki's three novels to date: *My Year of Meats* (1998), *All Over Creation* (2003), and *A Tale for the Time Being* (2013). Where Yamashita's narrative was an eerie mixture of Brazilian soap opera and magical realism, Ozeki's first novel assumes the format of a "Japanese docu-soap" (as the back cover reads) by blending the documentary/filmic narrative style with elements of a soap opera and an ecofeminist script.[36] In *My Year of Meats* such an agenda material-izes in its interrelated indictments of the meat industry and of sexist vio-lence against women,[37] while in *All Over Creation* Ozeki resorts to the melodrama typical of toxic discourse (Buell 1998) in order to denounce the spread of genetically modified organisms (GMOs) all over the world (see Wallace 2016, 93–122). In her first two novels, therefore, Ozeki concentrates on the materiality of our bodies and the materiality of our planet. In her third novel, however, the writer apparently changes course and attempts to tease out the complexities of both material and immaterial nature.

Ozeki's *My Year of Meats* describes the lives of two women, Jane and Akiko, as they become intermittently connected until they eventually meet. This is a healing encounter whereby transnational female bonding overcomes the patriarchal violence, both physical and psychological, that Akiko suffers at the hands of her husband. Sexuality and reproduction feature prominently in these women's lives, promoting an unexpected sisterly bond. For very different reasons and under totally different circumstances, both Jane and Akiko are denied their reproductive rights, either through the Diethylstilbestrol (DES) hormone that makes Jane initially barren or through sexist domestic violence in Akiko's case. Paradoxically, although international media corporations will also be accused of manipulation, it is globalized TV that serves as the umbilical cord between the two women.[38] In other words, it is the global media that erases the physical, cultural, and social distance between the laid-back Akiko, a housewife living in Japan, and the resolute Jane, a Japanese American filmmaker. It is thanks to Jane's work in the TV documentary series *My American Wife*, a program whose Japanese side is coordinated by Joichi, Akiko's husband, that the two women get to know of each other's existence.

At the same time that this symbolic sisterhood is gradually built, Jane starts suspecting the dangerous and foul games enacted by the very meat company that funds her documentary programs. The collusion of the all-powerful meat industry and the media corporations, as described in Ozeki's novel, renders the pernicious effects of globalization even more conspicuous: "Each episode of *My American Wife!* carried four attractive commercial spots for BEEF-EX. The strategy was 'to develop a powerful synergy between the commercials and the documentary vehicles, in order to stimulate consumer purchase motivation.' In other words, the commercials were to bleed into the documentaries and documentaries were to function as commercials" (Ozeki 1998, 41). As expected, the TNCs bear the imprint of American neo-colonialism, as both the TV and the beef company are United States based and intend to export not only their American products but also the genuine "American way of life." The exportation of American values takes place both as the underlying concept behind the culturally pregnant food metaphor and, in a more explicit way, in the precise images of American domestic bliss that are tailored to suit the company's conservative expectations.

Significantly, Jane's successful erosion of both power structures, food corporations and global media, begins with her gradual transformation of the image of the sweet all-American housewife cooking for her perfect family.[39] Jane starts following Joichi's injunction to "catch up [*sic*] healthy American wives with most delicious meats" (10). Joichi's "list of Desirable Things" excludes any visible flaws: "The BEEF-EX people are very strict. They don't want their meat to have a synergistic association with deformities. Like race. Or poverty. Or clubfeet" (57). As the representative of the company's interests, Joichi will not allow any departure from the prescribed script and recoils at the very thought of featuring an African American woman cooking chitterlings. As we can see in the following conversation between him and Akiko, Joichi construes Jane's attempt to film Miss Helen's African American dish as the first sign of the deterioration of his "perfect family" policy:

> The whole point is to show perfect families. We don't want families with flaws. And anyway, you should have seen the other family. ... Of all the stupid ideas, she [Jane] actually thought this black woman could be an American Wife!
>
> What was wrong with the black woman?
>
> ... You should have seen her family. First of all, they were extremely poor. Their accents were so uneducated that even with my level of English I could barely make out what they were saying. And that was just the beginning. The husband had terrible dentistry, gold teeth everywhere, and the wife just looked, well, badly dressed. Their house was not beautiful at all, and the food she cooked! Pig intestines! Entirely inappropriate
>
> It sounds ... different. (129)

However, in a gradual manner, Jane adds new, unexpected ingredients to the documentary recipe: a Mexican American immigrant family with a handicapped father; a couple with twelve children, ten of them adopted from different countries; and the last drop, a mixed-race homosexual couple, who turn out to be vegetarians. As each chapter introduces a new, provocative element, the unease among Jane's superiors mounts to the point of imperiling her job, until she decides to film a documentary on an apparently ideal family of meat producers. What is first welcome as a necessary return to the traditional, conservative script turns out to be just the opposite: the unethical practices that Jane's program eventually uncovers end up indicting the entire meat industry.[40]

In *All Over Creation*, Ozeki once more explores the politics of food production. If the focus of her first novel was the global meat industry, in her second Ozeki shifts her attention to the potato agribusiness.[41] While her debut novel concentrated on one type of "toxic food," hormone-treated meat, her second novel examines the insidious ways in which TNCs like Monsanto (evoked by a similar imaginary brand, Cynaco) forces GMOs on potato farmers, an imposition that not only endangers planetary biodiversity through bioengineered homogeneity but also becomes a real health hazard.[42] Thus, Ozeki's first two novels engage in "militant advocacy against certain practices of the agribusiness corporations that provide food for most Americans" while at the same time exploring "cultural hybridity and transnational belonging" (Heise 2008a, 395). A feminist agenda is also visible in both *My Year of Meats* and *All Over Creation*. The latter continues its focus on sexuality and reproduction rights, as the main character, Yumi, struggles with memories from her youth, including a teenage abortion.

What is significantly different in this second novel, however, is the tone. Even if tinged by melodrama, the mood in *All Over Creation* is far more somber than that encountered in *My Year of Meats*. Another distinguishing element in *All Over Creation* is Ozeki's inclusion of different models of ecological consciousness. On the one hand, we meet Yumi's old parents, whose conservative and conservationist attitude toward human embryos and plant seeds, respectively, is reminiscent of traditional notions of stewardship; on the other hand, we meet the Seeds of Resistance, a group of young environmental activists who indulge into some sort of new-age Gaia cult and whose ideology is indebted to more radical discourses, like that of deep ecology. While one would expect these two cohorts to be at odds with each other, the novel shows the Seeds joining forces with Yumi's father in their shared battle against Cynaco's practices. Such a remarkable alliance may respond to the author's conscious effort to expose the link between the need to preserve human life, ethnoracial diversity, and biological diversity.[43]

Some ecocritics have approached *All Over Creation* as a relevant attempt to grapple with the increasing perception of globally widespread toxicity that Ulrich Beck (1986) calls "risk society" (*Risikogesellschaft*), as Molly Wallace convincingly argues in her recent *Risk Criticism* (2016). Wallace chooses to read Ozeki's *All Over Creation* as a good example of Buell's toxic discourse. She claims that, with this narrative, toxic discourse has moved "into new territory, for if, as Buell says of the hazards represented in such discourse, 'the case has not yet been proven, at least to the satisfaction of the requisite authorities,' this is doubly true in the case of genetically modified foods, which are often in fact offered as the environmental solution to the problems of pesticide poisoning" (Wallace 2016, 99). Therefore, GMO novels like Ozeki's necessarily rely on "allegation rather than [on] proof" and, as Buell notes of toxic discourse, they often resort to "'moral melodrama' to persuade its audience of the reality of the hazard" (99).

Other ecocritics, like Heise, have focused on the intricate ways Ozeki intertwines issues of biological and cultural hybridity. In "Ecocriticism and the Transnational Turn in American Studies" (Heise 2008a, b), Heise notes that Ozeki's novel pits two agricultural models: "large-scale Idaho potato farmers, agribusiness corporations, genetically engineered crop varieties, and the global markets they command with the Fullers' mom-and-pop business, their small-scale cultivation, and dissemination of unaltered plant seeds" (2008a, 396). In building such a contrast, however, Ozeki contrives a metaphorical identification between humans and plants that can prove problematic. For Heise, Ozeki's deployment of "cultural diversity [...] as a substitute, complement, or metaphor for biological diversity" becomes "a way of endorsing the cultural encounters that globalization processes enable" while dismissing less palatable aspects of globalization (400).

Last but not least, *All Over Creation* has also been interpreted as an act of resistance against America's neoliberal, imperialist script. In her eloquent interrogation of the role that the nostalgia for nature has in American literature, Jennifer Ladino (2012) devotes a chapter to Ozeki's novel. Ladino reads *All Over Creation* as a deeply post-pastoral narrative that proves that "nostalgia does not always escort American hegemony across new frontiers" (198). In order to do so, Ozeki juxtaposes and compares social and biological diversity, a move that Ladino, in contrast to Heise, sees under a positive light (215–216). While Ladino also highlights the correspondence between cultural and biological hybrids, for "the nostalgia for natural 'hybrids' that fuels the book's criticism of agribusiness and its monocultures parallels the hybridity of human beings, and of human cultures," in her opinion, Ozeki's novel ultimately "underscores a celebration of multicultural human identity" and upholds the uncommodifiable nature of "nature" against the American neoliberal script (215, 219–220).

While both *My Year of Meats* and *All Over Creation* proffer an overt denunciation of the impact of transnational agribusiness corporations, Ozeki's third book, *A Tale for the Time Being* (2013), adopts a more oblique stance in its negotiation of environmental dangers. Although the author does address global environmental issues like sea pollution or climate change, she arguably uses such elements as anchoring points for the core of her narrative, which seems to revolve around more metaphysical matters, such as human suffering, the evanescence of life, and mortality. However, Ozeki's take on such classic literary topoi is rather unorthodox, in that she wittily combines traditionally antagonistic discourses such as science (quantum physics)[44] and religion (Zen Buddhism) in an attempt to understand global interdependence, indeterminacy, and the interaction of matter and consciousness.[45] I contend that *A Tale for the Time Being* engages in a much-needed "global cognitive mapping" (Jameson 1991, 54) of our late-capitalist world. In particular, in what follows I will try to demonstrate that Ozeki's book can be read *both* as an excellent illustration of David Harvey's materialist theory of time-space compression *and* as an interrogation of the limits of materialism. As we shall see next, in her choice of central tropes for her narratives, Ozeki has significantly moved from tangible meat and potatoes to both material and immaterial *gyres*.

3.2   Time-Space Beings: A Tale for a Globalized World[46]

Both the structure and the subject matter in *A Tale for the Time Being* bespeak interconnectedness. Ozeki's third novel, like her debut in fiction, *My Year of Meats*, is articulated around a bicontinental, trans-Pacific "conversation," this time between a Japanese teenager and an American writer living in British Columbia. However, on this particular occasion the two characters are separated not just by space—the Pacific Ocean—but also by time, since they are writing from different chronological sites, before and after the tsunami in Japan. *A Tale for the Time Being* braids two main narratives: the diary of a Japanese girl who is contemplating suicide, Nao, which opens the novel, and the response by Ruth, the person who finds Nao's journal some years later.[47] At first sight, Nao's first-person narrative seems mostly concerned with the teenager's sheer survival in a hostile social environment, that of the Japanese high school she attends after returning from California, where she has spent most of her childhood and early adolescence. On the other hand, Ruth's section of the book, intriguingly narrated in the third person, tackles both personal and global matters, including environmental issues.

The novel is significantly triggered by a fortuitous finding, a Hello Kitty Box, after a recent (un)natural disaster, Japan's earthquake and tsunami, and their anthropogenic amplification, Fukushima's nuclear crisis. From the very beginning Ruth refers to global environmental degradation, most notably the anthropogenic pollution of oceans (Ozeki 2013, 187). Her segment of the narrative "dialogue" starts by mentioning jetsam/flotsam—garbage in general, and plastic in particular—followed by the attendant description of planetary gyres and the "Great Garbage Patches." Both in her personal reflections and in the course of conversations with her partner, Oliver, and their few neighbors and friends, we find allusions to environmental problems like pollution and global warming (e.g. increasing presence of jellyfish in what used to be cold sea water), as well as discussions of the consequences of the earthquake and tsunami, especially Fukushima's nuclear crisis. These allusions are complemented by references to some of the quandaries currently encountered within the discourse of ecology, such as the distinction between native and invasive species, a question that comes up when Oliver's polemical "NeoEocene site" is described (120).[48]

This list of environmental worries may give the impression that Ruth, Ozeki's obvious *alter ego* in *A Tale for the Time Being*, is merely used by the author to lecture on and voice her own ideas about a given subject. Ozeki, however, is very much aware of the fact that she is a novelist rather than an apologist for a specific cause. In a recent interview (Ty 2013, 161), the author defends her position, claiming that, when writing her novels, she never has a specific agenda.[49] We can presume, then, that Ozeki has become more and more conscious of the danger of sounding too pedagogical or over-explanatory. In *A Tale for the Time Being*, Ozeki is judicious enough to foresee that pitfall and effectively dodge it. If she feels she needs to include scientific theories, she tries to make scientific exchanges relevant to the story or else moves them to appendices to the novel itself. As a result, the description of ocean gyres, for instance, is deftly integrated as the plausible explanation of why Nao's diary could have reached Ruth's hands when it did, and the theories of quantum mechanics, as we shall see, are not so much presented as erudite digressions—those appear in an appendix—as literally "fleshed" out in the form of a nonhuman animal. Equally interesting is the narrative strategy that surfaces on those occasions when characters other than Ruth seem about to engage in lengthy debates about technical and scientific matters. When this happens such discussions appear as secondary or background conversations, which Ruth only catches bits and pieces of, as when her friends discuss the transformation of plastic into everlasting "confetti"[50] pervading ocean water (93–94): when the explanations start to sound too much like a lecture, Ozeki, rather skillfully, has Ruth interrupt her friends by changing the subject.

In addition to focusing on the materiality of the world and voicing environmental concerns, *A Tale for the Time Being* investigates more immaterial concerns, such as postmodern *angst* and human mortality. The existential dilemmas in the novel initially emerge from personal situations: Ruth's writer's block, her mother's Alzheimer's, and Nao's suicidal drive, partly (but not exclusively) caused by the harrowing experience of being bullied. However, these personal troubles soon expand to include larger issues of an immaterial, philosophical, or eschatological nature, including the anxiety of living in a postmodern, globalized world. In particular, in its exploration of the ways in which the new transnational and networked spaces affect the very notion and constitution of (human) nature, *A Tale for the Time Being* chooses to focus on the changing time-space paradigms and their impact on our lives.

In the 1980s and early 1990s critics like Fredric Jameson (1984, 1991) and David Harvey (1990) started to both announce and scrutinize the advent of postmodernism, which had become "the cultural dominant of the logic of late capitalism," the "third great original expansion of capitalism around the globe" (Jameson 1991, 46, 49), what we now know as contemporary globalization. In *The Condition of Postmodernity*, Harvey tried to examine "the material practices from which our conceptions of time and space flow."[51] At the time when he was writing, the critic saw that previous static models were no longer useful. By the 1970s the Fordist-Keynesian paradigm, broadly accepted in the capitalist world since the end of WW2, had started to erode, and a new model had come to replace old Fordism, a model that Harvey called the regime of "flexible accumulation," a new system that "exploit[ed] a wide range of seemingly contingent geographical circumstances, and reconstitute[d] them as structured internal elements of its own encompassing logic" (1990, 294), allowing for a drastic reduction of "turnover time" in production, circulation, and consumption (156, 291). Such time-space compression would change everything.

This late form of capitalism was both postindustrial and globalized. Firstly, from their Western perspective, these scholars claimed that the new economic system emerging in the last decades of the twentieth century was clearly postindustrial, as the "rapid contraction in manufacturing employment after 1972" in capitalist Europe and in the United States proved (Harvey 156–7; this, of course, disregarded the fact that other actors, like China, were becoming industrialized). Concomitantly, in those years there was a "shift away from the consumption of goods and into the consumption of services," including what Harvey called "distractions" (285). Therefore, in contrast to previous models, late or "advanced" capitalism did not concern itself so much with materiality as with immaterial services, financial products, entertainment, and information and communication technologies. After all, the systems of production in our current economic system, as Phillips eloquently puts it, have become "primarily and splendidly electronic": "Capitalism's boldest endeavors, no longer involve the extraction of raw stuff from the earth, but endless recycling. However, it isn't the recycling of paper, plastic, glass, and other non-quite-consumables that interests venture capitalists, … but the elliptical orbits of credit, debt, imagery, and information" (2003, 26).

Secondly, historians and theorists concurred that the new economic system was a globalized form of capitalism. Although a global market had existed for much of the twentieth century, when Fordism was the most popular form of capitalist production, the effects of globalization were more crucial in the new system of flexible accumulation that emerged in the late twentieth century. Time and space seemed to be abolished to all effects, as far as banking and finance were concerned. With the end of the Bretton Woods agreement and the advent of "floating exchange rates" (Harvey 1990, 296; see Stiglitz 2017), the global financial system was reorganized into "a single world market for money and credit" (160–161). Having managed to escape the usual "constraints of time and space that normally pin down material activities of production and space" (164), money had finally become "stateless" (163) and "dematerialized" (297). If capitalism had always longed for the "annihilation of space through time" (293), the new regime of flexible accumulation, facilitated by technological innovation, seemed to make such annihilation possible.[52]

It is hard to dispute the fact that, in our globalized world, the final "annihilation of space through time" seems nearer than ever: distances have been bridged by improved means of transport, especially air travel, but also, and more crucially, by the revolution in information and communication technologies (ICTs). When Jameson, McLuhan and Powers (1989), or Harvey first theorized the impact of mass media and television upon our lives, in the 1980s and early 1990s, they could barely imagine the ICT revolution that would take place at the turn of the century, a revolution that has led intellectuals like Manuel Castells (1996, 1997, 1998) to dub this era the digital "information age." Theoretically at least, the new ICTs, most notably the Internet, have made it possible for people from different parts of the world to communicate, usually in an instantaneous way, although, as we shall see in Ozeki's novel, global interconnectedness does not always mean genuine communication.

Like much of Yamashita's fiction, Ozeki's last book belongs to that group of recent American novels that Rachel Adams has dubbed "American literary globalism" (2001) and which Min Hyoung Song deftly reviews in "Becoming Planetary" (2011). In *A Tale for the Time Being*, Ozeki scrutinizes the current dynamics of global interdependence in intriguing ways, combining new images with traditional ones. Echoing Yamashita's *Through the Arc of the Rain Forest*, whose narrating ball resembled a tiny planet, or at least a satellite, in her novel Ozeki also chose to introduce elements resembling and symbolizing small planets: among the flotsam found on the beaches of British Columbia, people find "old Japanese fishing floats that had detached from nets across the Pacific …, *murky globes* blown from thick tinted glass. They were beautiful, like *escaped worlds*" (33; emphasis added). Such flotsam and jetsam are at the very origin of the novel, whose genesis can be traced back to the encounter with the debris from the 2011 Japanese earthquake and tsunami. When this debris started "to wash up on the West Coast" some time after the earthquake, for Ozeki this was "such a visual and material enactment—evidence that *the planet is really small,* and we are all *radically interconnected*" (Ty 162; emphasis added) that she needed to capture that epiphany in a novel.

Because of its formal characteristics, content matter, and structure, *A Tale for the Time Being* constitutes an excellent attempt at capturing the time-space compression brought about by contemporary globalization. In Ozeki's novel the time-space compression that Harvey deemed typical of our era is conspicuous enough. It is visible in the ubiquitous presence of information and communication technologies—the main facilitator of time-space compression in our days—in the very structure of the novel, with its alternation of different narrative agents and chronotopes,[53] and, more importantly, in the astute deployment of different planetary-temporal images that reinforce the intuition that space has indeed been "annihilated" in the era of "advanced," globalized capitalism.

As might be expected of contemporary "global citizens," the main characters in *A Tale for the Time Being* make constant use of information and communication technologies. Ruth, living as she does in a remote island, is dependent on email for communication and on global media and the Internet for information; Nao is both an avid user of ICTs and a victim of cyber-bullying[54]; her father is an ICT specialist who loses his job after the dot.com bubble bursts; even Jiko, the ancient Buddhist nun who comes from a "prewired world," learns to text in order to communicate with her great-grandniece Nao. However, it is not only messages and information that travel in Ozeki's novel but also species and objects (commodities, garbage) that move or drift from one continent to another, in such a way that it becomes difficult to tell where they are from, since, like the Matacão in Yamashita's novel, they are all enmeshed in more-than-local processes. Thus, Japanese crows become a common sighting on the West Coast of North America. Debris from the tsunami, like the Hello Kitty Box where Nao has hidden her diary, also starts arriving at Ruth's shores. When Ruth learns that even the particular type of oysters now harvested in British Columbia originates in Japan (precisely in the district, Miyagi, where Jiko is from), Ruth cannot but feel "the wide Pacific Ocean suddenly *shrink* just a little" (187; emphasis added). We are immediately reminded of Harvey's description of postmodern compression, whereby "space appears to shrink" (1990, 240).

The same shrinking of the world is symbolically conjured up when Oliver tells Ruth that, as a consequence of the 2011 earthquake, Japan has actually moved closer to North America, albeit on a negligible scale. In a reversal of the prehistoric continental drift, the narrative evokes the coming together of both landmasses: when Ruth wonders whether they "should go to Japan," Oliver retorts "Maybe we don't have to, since Japan is coming here" (202). On hearing this, many readers will remember Japanese crows and Hello Kitty boxes; however, what scientifically minded Oliver means is that the 2011 earthquake has actually "moved the coast of Japan closer to us" (202), in spatial, geographic terms. On this occasion, Harvey's theory of time-space compression becomes literal, since the earth rotation is slightly altered: the earthquake, says Oliver, "caused the planet's mass to shift closer to the core, which made the earth spin faster. The increase in the speed of rotation shortened the length of the day" (203). As a result, Oliver continues, "*[o]ur days are shorter* now" (203). Even though it is just an unperceivable fraction of time that is "lost," the symbolic value of this realization should not be underestimated.

Although Ozeki hints at time-space compression in the aforementioned scenes, it is through two intriguing leitmotifs that she openly scrutinizes the quandaries of contemporary time-space paradigms: the polyvalent symbol of the gyre and the trope of the "message in the bottle." As mentioned before, Ruth learns of Nao's existence when she finds the girl's diary in a plastic-wrapped box that has drifted all the way from Japan. Even though both Nao and Ruth frequently use the Internet and other ICTs, the girl's experiences, especially the cyber-bullying she has had to endure since her arrival in Japan, make her distrust these technologies. Nao's father is the one who puts it more bluntly: "The Internet's a toilet bowl. ... We never thought it would turn out like this" (351). That is why, in order to send out her existential SOS, Nao, technologically savvy though she is, resorts to the old-fashioned "message in a bottle" instead of the apparent global interconnectivity of the Internet. Ozeki's deployment of such a premodern means of communication in our ICT-saturated world echoes Yamashita's paradoxical choice of pretechnological pigeons to explore the extent of human interconnectedness in the age of globalization. Likewise, in *A Tale for the Time Being*, Nao's choice of such a primitive method underscores her lack of confidence in the ICTs' ability to facilitate real human communication. When Nao gives a more precise description of her strategy, the differences between the various forms of human contact become more evident: her floating lunchbox resembles "a message in a bottle, cast out onto the ocean of time and space. Totally personal, and *real*, too, right out of old Jiko's and Marcel's prewired world. It's the opposite of a blog. It's an antiblog, because it's meant for only one special person" (26, emphasis added). The narrator underlines the fact that this "message in a lunchbox" is tangible, "real," in implicit contrast to the virtual world of the Internet, where one can have many "friends" whose friendship is far from genuine. In the description of her chosen method, Nao highlights the sense of touch, which makes all the difference in human communication:

> I will write down everything I know about Jiko's life in Marcel's book [À la recherche du temps perdu], and when I'm done, I'll just leave it somewhere, and you will find it!
> How cool is that? It feels like I'm reaching forward through time to touch you, and now that you've found it, you're reaching back to touch me! (26; italics in the original)

Ruth will also feel that physical connection at regular intervals, as when she smells her diary to know more about Nao (37–38) or when she fingers "the blunt edges of the diary" and wonders whether Nao had also "worried this corner between her fingertips" (34). Such literal touch echoes the more important realization that Nao's diary has "touched" or moved Ruth in deep, disturbing ways.

Shifting our focus to the materiality of Nao's "message in a bottle," we cannot help but notice that, in this case, the anonymous, crystal-clear bottle of old times has been replaced by a less transparent container. The schoolgirl lunchbox, deceptively innocent looking, contains Nao's diary, her uncle Haruki's letters, and his old watch. Nao's chosen container bears a commercial label, Hello Kitty, a TNC brand that conjures up the new reality of global trade. Furthermore, this time the watertight container is not a crystal bottle, with its cork; instead, it is ecologically insidious plastic that wraps up the box and keeps it from getting wet as it floats on the ocean. At the same time, the lunchbox adrift on the ocean not only carries Nao's message, but also functions metonymically as a carrier of Nao herself. The intimate connection between the floating lunchbox and Nao is confirmed when we read her description of the word *rōnin*, a type of Japanese warrior whom Nao comes to identify with: the Japanese ideogram for *rōnin* is made up of two characters, one "for wave" and one "for person, which is pretty much how I feel, like a little *wave person*, floating around on the stormy sea of life" (42; emphasis added). If the lunchbox and the treasure it contains serves as a *metonymic* reference to Nao, since it is "part of her" or at least of her most personal belongings, the "wave person, floating around on the stormy sea of life" activates the reader's *metaphorical* muscles and immediately brings to mind the floating box. The analogy is obvious: both Nao and her lunchbox need to be "rescued" and "touched" by the right person, in this case Ruth.

Therefore, Ozeki's *A Tale for the Time Being*, with its profusion of letters, diaries, "restless pixels" (227), emails, texting, and other forms of communication, can be approached as an insightful exploration of the limits of human communication and, more specifically, of the ways in which contemporary globalization has both facilitated and impeded genuine communication. We have just seen how the tangible, real ocean both makes possible and prevents communication between Nao and Ruth. At the same time, the sea provides the narrator with an effective metaphor—"the ocean of time and space" (26)—that not only underscores the (initial) distance between Nao and Ruth but also suggests that time and space are deeply imbricated and entangled with each other. Intriguingly enough, in order to describe *both* the spatial pattern of ocean pollution *and* the time-space compression brought about by ICTs, Ozeki resorts to the same image: the *gyre*. Planetary gyres are first mentioned in connection with the Great Garbage Patches made up of plastic. Oliver explains that bags, "bottles, Styrofoam, take-out food containers, disposable razors, industrial waste," in a word, "[a]nything we throw away that floats" drifts and finds its way into these ocean gyres (36). Such anthropogenic islands prove highly resilient, because plastic "never biodegrades. It gets churned around in the gyre and ground down into particles," and, in that "granular state, it hangs around forever" (93). As might be expected, such "plastic confetti" is ingested by all sorts of ocean species and, through the trophic chain, it ends up in our bodies (93–94). In the long run, plastic toxicity may be almost as insidious as radioactivity, even if the half-life of radium is significantly higher.

If spatial, ocean gyres facilitate the accumulation of (tangible) garbage, "temporal gyres" invoke the (less tangible) time-space compression that inheres in global capitalism, mostly via information and communication technologies. At the same time, both geo-drift and "info-drift" are indicative of our "throwaway society" (Toffler), a society where everything rapidly "decays" and is discarded. The new developments in "advanced" capitalism have built a society predicated on ephemerality and impermanence; an obvious consequence of time-space compression is an overwhelming sense of transience, of "instant obsolescence" (286), as Harvey already phrased it in 1990. Ozeki finds in the vertiginous *gyre* an appropriate figure to signify the "sensory overload" brought about by ICTs' "instantaneous communicability" (Harvey 289, 288) and the concomitant sense of ephemerality that seems to pervade everything, from plastic products to the news:

> What is the half-life of information? Does its rate of *decay* correlate with the medium that conveys it? ...
>
> Does the half-life of information correlate with our attention? *Is the Internet a kind of temporal gyre, sucking up stories, like geodrift, into its orbit?* What is its gyre memory? How do we measure the half-life of its drift? (Ozeki 114; emphasis added)

This pairing of spatial and temporal gyres necessarily evokes time-space compression, ecological interdependence, and global connectivity, while it also emphasizes the shared toxicity of physical and digital worlds. The plastic that concentrates on the great garbage patches, decays, and becomes everlasting "confetti" pervading and polluting our ocean water (93–94) is as toxic as the (mis)information traveling, "decaying," and (un)mercifully reverting to its pixel-confetti nature. Ozeki's novel conjures up contemporary time-space compression at the same time that it shows its less pleasant underside. In a world saturated with the information and images that global mediascapes and ICTs constantly feed us, the trope of the gyre allows Ozeki to highlight the toxic implications of both geo-drift and "info-drift," while Nao's pretechnological means of communication, harking back to the "message in a bottle" of old times, offers a somewhat naive but still intriguing alternative. In order to reach and touch a significant other, Nao chooses a premodern artifact that, precisely because of its limitations, allows genuine human connectedness across time and space.

3.3   Buddhism and Quantum Physics: A Tale of the Immaterial World[55]

The realization of an increasing economic interdependence in contemporary globalization has been accompanied by a similar awareness of the planet's ecological interdependence.[56] Although Ozeki hints at both the interdependent planet and the intensified connectivity of globalization, she is concerned with a much deeper type of connectedness, which, as the writer confesses in her 2013 interview with Eleanor Ty, finds its roots in Zen Buddhism: "much of what I am writing about in [*A Tale for the Time Being*] is informed by basic Buddhist principles—of interdependence, impermanence, interconnectedness" (161). It might be argued that, by turning a Buddhist nun, Jiko, into a feminist heroine and a mentor for the victimized Nao, and by highlighting her teachings, this novel implicitly endorses a Buddhist worldview.[57] In other words, what Ozeki is trying to complement is the materialist understanding of connections across space and time with the spiritual dimension of "time beings." In fact, Part 3 of the book opens with a quotation from *Shōbōgenzō*, written by Zen master Dōgen centuries ago, a quotation that introduces the concept of *uzi* or time being: "everything in the entire universe is intimately linked with each other as moments in time, continuous and separate" (30). Expressed in this way, Zen Buddhist beliefs seem germane to the central ecological principle of interdependence.[58]

Even more interesting is the fact that, in her exploration of interconnectedness, Ozeki tries to combine two perspectives usually seen as antagonistic: religious and scientific discourses. Arguably, Ruth and Oliver embody two different frames of mind, artistic and scientific, respectively. Although Ruth seems quite a rationalist skeptic herself, when she becomes engrossed in Nao's diary and learns about the Buddhist nun, she starts to move away from Western skepticism and seems more open to a spiritual understanding of life.[59] In contrast, Oliver seems to stand for scientific rationality. Even before we see him "flipping through the latest issue of

New Science magazine" (66) or lecturing on ocean gyres and garbage patches (13–14; 35–36), Oliver is shown to have a curious scientific mind, always dissecting and trying to understand phenomena from a rational point of view, an activity graphically described as "forensic unpeeling" (9). More surprisingly, a nonhuman character, Ruth and Oliver's cat, seems to be associated with scientific discourse as well. In contrast to the childish echoes of the other cat in the novel (Hello Kitty), the name of this cat, "Schrödinger," has obvious scientific resonance: its name immediately evokes quantum mechanics, especially the principle of indeterminacy illustrated with the metaphor known as Schrödinger's cat (62, 396, appendix E). Erwin Schrödinger's imaginary experiment consists of a steel chamber where a cat has been introduced. In such a box there is a device that can release some poisonous substance, but that mechanism works in an aleatory way: it may or may not poison the cat. After some time, the external observers cannot know for certain whether the cat inside the chamber is dead or not, so they have to work with both hypotheses at the same time.[60]

How exactly does Schrödinger's imaginary experiment intervene in Ozeki's *A Tale for the Time Being*, beyond providing a name for the couple's cat? I would argue that the author tries to give "flesh" to the thought experiment by having a real cat vanish and keeping his state unknown for a good part of the novel. When Schrödinger, or Pesto, as they often call him, is nowhere to be found, Oliver starts to worry, especially because rumors abound of raccoons attacking dogs in the neighborhood. Significantly enough, when Schrödinger finally reappears, more dead than alive, Oliver resents not so much the pain that the cat has suffered as the uncertainty that plagued him during those days: "That's what was driving me crazy. Not knowing where he'd gone, or if he was alive or dead [as in Schrödinger's experiment]. But at least now we know. ... There's nothing worse than not knowing" (381). In other words, Oliver longs for scientific certainties and does not seem comfortable with the glimpse of indeterminacy that can derive from Schrödinger's imaginary experiment. If the parallel with quantum physics was not clear enough, at the end of the novel Ruth explicitly mentions that Pesto/Schrödinger brings to mind the superposition of states in quantum mechanics. Oliver and Ruth are visiting the convalescent cat, and when she is about to caress his head, she asks:

> That cat of Schrödinger's [...] reminds me of you. What quantum state were you in when you were *hiding in the box* in the basement?
> "Oh," [Oliver] said. "That. Definitely *smeared*. Half-dead and half-alive. But if you'd found me, I would have died, for sure." (400; emphasis added)

Ruth and Oliver's exchange is a direct reference to Schrödinger's paradox, which ends when the observer opens the box or chamber and discovers whether the cat is dead or alive. Until that moment both possibilities coexist. The same superposed states exist in the universe ruled by quantum mechanics. As Oliver explains, "the particles exist in superposition only as long as no one is looking. … Until that moment of observation, there's only an *array of possibilities*, ergo, the cat exists in this so-called smeared state of being" (397; emphasis added). According to this interpretation, observation puts an end to the indeterminacy. Yet Oliver hastens to add, another physicist, Hugh Everett, came up with a different solution and claimed that, even after observation, "the superposed quantum system persists"; what happens is that "when it is observed, it branches. The cat isn't either dead or alive. It's *both* dead *and* alive, only now it exists as two cats in two different worlds" (397).[61] If that were true, Oliver notes, "we're now in a world where Pesto is alive, but there's another world where he was killed and eaten by those dastardly coons" (399).

The recurrent allusion to the temporal uncertainty about the real cat, echoing as it does Schrödinger's paradox, ultimately helps to erode our most basic human certainties. In the context of the novel, the fact that for some time we don't know if Pesto/Schrödinger is dead or alive highlights the uncertainty about Nao's life. At the end of the novel, there is no way of knowing whether Nao is still alive. Not only that, but when Ruth remembers Nao's initial words addressed at her anonymous reader, *Together we'll make magic*, Ruth starts entertaining doubts about her own existence. She can no longer be sure whether Nao has called her into being or the other way around:

> Who had *conjured* whom?
> […] Was she the *dream*? Was Nao the one writing her into being? […] She'd never had any cause to doubt her senses. Her empirical experience of herself, as a fully embodied being who persisted in a real world of her remembering, seemed trustworthy enough, but now in the dark, at four in the morning, she wasn't so sure. (392; emphasis added)[62]

Here Ruth draws together different approaches to "reality": the scientific empiricism explicitly mentioned and nonscientific discourses suggested by words like "conjuring," "dreaming," or "creating." To this we should add the religious discourse of Zen Buddhism. Ozeki has recently maintained that Buddhist principles and praxis seem to adapt better to the postmodern subjectivity. It comes as no surprise, then, that some characters in the novel should embrace, or at least value, a Buddhist understanding of the impermanence and uncertainty of life. Such uncertainty and indeterminacy are not too far from the theory of multiple worlds deriving from quantum mechanics. Ozeki herself (both in her interviews and in the appendices to the novel) links the principles illustrated by Schrödinger's thought experiment with Buddhist non-duality: "We think of these states as either/or, but what if they're not? What if they're just *and & and* and *and & and*? This ties into another Buddhist notion, that of non-duality. I think all of these things—non-duality, impermanence, interdependence, no abiding self—all of these themes come from Buddhist teaching" (Ty 2013, 166).[63] The Buddhist belief in non-duality seems to coincide with the theory of multiple worlds: "if you buy the many-worlds interpretation of quantum mechanics, then everything that's possible will happen, or perhaps already has" (Ozeki 2013, 395). We may even wonder if death is possible at all "in a universe of many worlds" since, by their very existence, they "guarantee a kind of immortality" (400).

It is from this philosophical perspective that we should try and understand the ending of the novel. Like Oliver and Ruth, when we readers reach the last pages of the book, we may wonder "what happened to Nao during the tsunami," because, as Ty points out in her interview with Ozeki, "that is never really answered for us" (2013, 169). The novelist recognizes that she does not "really know what happened" either: "a lot of people just simply vanish. So, who knows?" (169). When Ty suggests that the fact that the cat did reappear in the end may be indicative of a similar fate in Nao's case,[64] Ozeki resists the temptation to provide closure and leaves us all in the dark.

"Not knowing is hard," the narrator declares toward the end of the novel (Ozeki 2013, 400). Like Oliver, who suffered not knowing what had happened to Pesto, Ruth detests uncertainty: "I'd much rather know, but then again, *not-knowing keeps all the possibilities open. It keeps all the worlds alive*" (402). Hesitating between the two options, Ruth asks scientifically minded Oliver whether he believes "there are other worlds where. ... Where no one dies in the earthquake and tsunami? Where Nao is alive and well. ... Where there are no leaking nuclear reactors or garbage patches in the sea?," to what Oliver replies: "There's no way of knowing" (399). Apparently, then, uncertainty wins the day.

As the foregoing analysis demonstrates, in her attempt to understand indeterminacy and interdependence, Ozeki has chosen to combine traditionally antagonistic discourses such as science and religion. Simultaneously drawing from Zen Buddhism and quantum physics, *A Tale for the Time Being* teases out the complex material and immaterial bonds between all organisms and particles. The novel does not shy away from paradox or mystery; on the contrary, it shows how both science and religion can offer ways to coexist with indeterminacy. Just as quantum physics theorists defend the "unbroken wholeness" of material nature (Bohm 1980, xvii–xviii), spiritual traditions like Zen Buddhism supplement immanence with transcendence in their exploration of interdependence. As both Kate Rigby (2014) and Greta Gaard (2014) have convincingly argued, even materialist schools of ecocriticism would benefit from such a strategic alliance (Rigby 2014, 284).[65] In *A Tale for the Time Being*, the juxtaposition of the religious and scientific discourses is a conscious move on the part of the author to highlight the fact that one does not have to choose between one or the other, that it is possible to reconcile both, that, in fact, it may be in our own interest to do so if both can help us save our planet. This realization is borne out by an apparently insignificant image. One night, as Ruth and Oliver lie in bed, reading and wondering about the many shadows and uncertainties surrounding Nao and her diary, Ruth notices the two types of light they are using: "Oil lamps and LEDs. The old technologies and the new, collapsing time into a paradoxical present" (344). In this scene, Ozeki is not merely describing the conjunction of old and new technologies, but hinting at the two discourses she has intertwined throughout the entire novel.[66] With the juxtaposition of oil lamps and LEDs, she is symbolically juxtaposing old and new ways of "throwing light" into our transnatural world.

4   The Many Layers of the Ozone Layer, or How to Reconcile the Material and the Immaterial

It is not language that has a hole in its ozone layer. (Soper, *What Is Nature?*)

In my earlier exploration of the current state of ecocriticism, as I addressed the recent interrogation of the nature-culture divide (the "transnatural" challenge), I observed that some ecocritics resent what they consider as the excessive emphasis on the constructed nature of nature. They are critical of the unquestioned credibility and reputation bestowed upon the linguistic turn in theory, which seems to reduce nature to little more than an idea. While we might contend that nature is linguistically and culturally mediated, it is equally undeniable that nature also has an extralinguistic reality. To use Kate Soper's famous words, "it is not language that has a hole in its ozone layer; and the 'real' thing continues to be polluted and degraded even as we refine our deconstructive insights at the level of the signifier" (1998, 168).[67]

To Soper's cautionary maxim that it is not language but the very real, atmospheric ozone layer that has a hole, cultural theorists like Latour would retort that such a "hole is too social and too narrated to be truly natural" (1993, 6).[68] To complicate matters, the ecocritical claim to realism more often than not translates into a blanket refusal to acknowledge what literary theory has achieved, namely, questioning the very ability of literature to represent something beyond it. While it remains important to invoke "our capacity for intimate acquaintance with nature," despite some theorists' insistence on its "entire otherness," Phillips also reminds us that we can no longer entertain the "hope that there is some beyond of literature, call it nature or wilderness or ecological community or ecosystem or environment, where deliverance from the constraints of culture, particularly that constraint known as 'theory,' might be found" (2003, 40).

At first sight, therefore, the task of reconciling the study and protection of "nature" with the antifoundationalist or antiessentialist thrust of poststructuralist theory seems rather difficult, if not impossible.[69] However, as Phillips demonstrates, "the choice between theory and nature is a false one, since neither comes to us with its pristine character intact" (2003, 40). Rather than positing nature and culture, or materiality and immateriality, as two realms perpetually at loggerheads with each other, we should try and reconcile both. When navigating these troubled waters, we may discover that, as happens in the estuaries, it is often hard to discern where the fresh water ends and the ocean water begins. The ozone layer is both natural and human: human because, as Latour rightly claims, it is inescapably narrated—and also because its dangerously growing hole was clearly anthropogenic. The ozone layer is both immaterial, because culturally mediated, and material, because it was its disintegrating materiality that sounded all the alarms and made us react by restricting the use of damaging CFCs.

The preceding analysis of Ozeki's and Yamashita's novels shows them to be amenable to an ecocritical reading that focuses on materiality, on "real" environmental degradation and insidious toxicity. These books engage in "a truly materialist version—ecologically rather than economically based— of 'cognitive mapping,' which must entail wiping our fingerprints off the landscape as we redraw the maps in our minds" (Phillips 1994, 219). At the same time, however, I contend that Ozeki's latest novel, *A Tale for the Time Being*, both enacts that "materialist cognitive mapping" and goes beyond materialism by bringing in the immaterial. Ozeki's narrative not only textualizes Harvey's time-space compression but also reconciles scientific discourses of matter with religious discourses of the immaterial. For material ecocritics openly indebted to immanent philosophy (Iovino and Oppermann 2014a, 3), this may seem an irresoluble *aporia*; yet Ozeki seems at ease with the apparent contradiction of embracing both immanent and transcendent traditions. While *A Tale for the Time Being* is very much concerned with the materiality of the oceans, of teenagers' bodies, of lost cats, it is also deeply entangled with the immateriality of spirituality. The fact that Ozeki manages to combine and reconcile the different philosophical, theological, and scientific theories of indeterminacy, impermanence, and interdependence in just one novel should not surprise us, because, as the author declared in a recent interview (Flood 2014), fiction is both a mirror and a shaper of reality, and "reality, like light, exists not only as a single point or particle, but also as an array of possibilities."

NOTES

1. In *I Hotel*, one of her most ambitious novels, Karen Tei Yamashita asks the resonant question: "Asian America (where's that?)" (2010, 230). See the website of the Asian American Writers' Workshop (AAWW), for an ongoing discussion of nature of Asian American literature: http://aaww.org/curation/page-turners-asian-american-literature-in-the-21st-century
2. In much the same vein, Stan Yogi coincided in the difficulty of finding a label for Yamashita's *Through the Arc of the Rain Forest*, at the same time that he claimed that the novel constituted a landmark in the Japanese American literary tradition, signaling as it did "a movement away from the treatment of specifically Japanese American characters toward a stunning blend of genres and characters" (1997, 148).
3. Since the 1990s a rising number of studies have focused on the shifting nature of Asian American literature. Lisa Lowe's "Heterogeneity, Hybridity, Multiplicity" (1991; cf. *Immigrant Acts* 1996) and Sau-ling Wong's "Denationalization Reconsidered" (1995) were the first articles to seriously tackle the tensions between the old monolithic-nationalist and the new transnational-diasporic paradigms in Asian American studies. The late 1990s saw a proliferation of books dealing with the topic, among them King-kok Cheung's *An Interethnic Companion to Asian American Literature* (1997), David Leiwei Li's *Imagining the Nation* (1998), Sheng-

Mei Ma's *Immigrant Subjectivities in Asian American and Asian Diaspora Literatures* (1998), and David Palumbo-Liu's *Asian/American: Historical Crossings of a Racial Frontier* (1999). These publications have been complemented by critical work appearing in the twenty-first century (Chuh and Shimakawa 2001; Nguyen 2002; Davis and Ludwig 2002; Simal and Marino 2004; Sohn et al. 2010; Simal 2013; or Parikh and Kim 2015).

4. For recent ecocritical work on the impact of the Anthropocene paradigm, see Clark (2014, 79–81), Heise (2016, 85–86, 204–205), Wallace (2016, 192), and Ortiz-Robles (2016, 6–7).

5. She elaborates on this vision of the "pervasively domesticated" nature in *Imagining Extinction* (2016), where she calls for "a new kind of environmentalism" that accounts for the reality of tamed, "post-wild" nature (12, 208).

6. As Dana Phillips convincingly claims, it was precisely the ecological crisis that triggered the emergence of ecocriticism itself: "Without environmental crisis ... there might be no 'environmental imagination.' At best, there would be only a very attenuated one. Nor might there be ecologists struggling to understand and repair the mechanisms of a damaged natural world. ... There is considerable irony in the fact that in order to begin to understand nature, we had first to alter it for the worse" (1999, 598).

7. An earlier, shorter version of the following discussion appeared in Simal's "The Junkyard in the Jungle," published in *The Journal of Transnational American Studies* in 2010. I thank the journal editors for kindly allowing me to reprint that material.

8. Not only the "nature of nature" but also its apparent uniqueness has been problematized in recent criticism. In the polemical *War of the Worlds* (2002), Bruno Latour maintains that the emergence of a postmodern "multinaturalism" (Viveiros de Castro) has already replaced the "mononaturalism" that had provided the bedrock for modernist confidence (20–21).

9. My understanding of the term "global" coincides with Robert Marzec's definition: "the complex of contemporary transnational forces of capital and culture governed by an endeavor to homogenize and reduce difference and distance, an evolving network first laid down in the eighteenth, nineteenth centuries with the development of the Dutch, French, and specifically British Empires" (2007, 25). For an overview of the different positions vis-à-vis globalization, see Scholte (2005), Jay (2010), Goyal (2017), and Simal (2018).

10. As we saw in Chap. 4, bioregionalism is far from dead, and some of the earlier ecocritics are rather skeptical about the effectiveness of global rather than local engagements with the natural environment (Berry 1989; Murphy 2009). Contrary to the positions held in earlier analyses of *Through the Arc of the Rain Forest* (Wallace 2000; Heise 2004; Simal 2010), John

Gamber has recently argued that Yamashita's novel traces Kazumasa's quest for a real home, which he ultimately finds not in Brazilian cities nor in the globalized Matacão, but in the "local" and rural location where he finally settles (2018, 43).

11. See Iovino and Oppermann's *Material Ecocriticism* (2014a). To the transnational and material(ist) turns, some ecocritics would add another new trend known as the "animal turn" (Weil 2010). See Chap. 1 for more information about the field of Animal Studies.

12. One example of this material turn in ecocriticism would be Nancy Tuana's viscous porosity (2008) or Stacy Alaimo's trans-corporeality (2010), two concepts discussed in earlier chapters), because of their emphasis on the interaction between the human bodies and the surrounding matter. Gamber's theorizations of positive pollutions (2012, 7–8, 122) are also relevant here.

13. In 1996 Jonathan Bate coined the term "Global Warming criticism" (436) in response to what he saw as a new paradigm and argued that the complexity of the weather "challenges the moderns' separation of culture from nature," exposing their "inextricability" (1996, 439; see Gamber 2012, 1, 15).

14. Among such literary topoi, many inherited from earlier gothic and pastoral traditions, Buell mentions the enlistment of "totalizing images of a world without refuge from toxic penetration," "images of community disruption" and "pastoral betrayal," or the reenactment of unequal David-Goliath battles (1998, 648–655). In *Positive Pollutions and Cultural Toxins* (2012), Gamber takes a different stance: while the critic is very much aware that the concept of toxicity has to be handled with care (6–7), he argues that it can be positively wielded in order to "represent a multitude of transgressive mixings that might be historically coded as negative, but which are demonstrated to be anything but" (7).

15. In the prelude we saw a clear example of toxic riskscapes in Mukherjee's *Jasmine* (1989), with its unnatural, nuclear plants. In addition, novels like DeLillo's *White Noise* and Updike's *Rabbit at Rest* underscored the fact that pollution had inexorably changed our experience of the planet itself "as primal home": this is "a generation poised on the precipice of an epistemic rupture"—between knowing the earth as "the landforms, flora and fauna which are the home in which life is set and knowing the earth as toxic riskscape" (Deitering 1992, 200).

16. Contrary to what might seem at first sight, Gamber claims, *Tropic of Orange* is as much about ecological concerns as *Through the Arc of the Rain Forest*, even though the former's "urban setting might preclude some readers from considering it an environmentalist text" (2012, 154). For an ecocritical analysis of *Tropic of Orange*, see Gamber (2012, 120–154).

17. Parts of the analysis of Yamashita's novel that I carry out in the following pages originally appeared in the *Journal of Transnational American Studies*

(Simal 2010). I thank the journal editors for kindly allowing me to reprint that material here. I also want to acknowledge the feedback I received at the International American Studies Association (IASA) Conference held in Lisbon in 2007, where I presented the paper "Meat or Plastic?: Glocalizing the Ecological Crisis in Recent Japanese American Fiction," a comparative study of Yamashita's *Through the Arc of the Rain Forest* and Ozeki's *My Year of Meats*, which would be the seed of this chapter.

A recent valuable addition to the scholarship on Yamashita's work is a collection of articles edited by A. Robert Lee. Some of the chapters in Lee's *Karen Tei Yamashita: Fictions of Magic and Memory* (2018) focus on *Through the Arc of the Rain Forest* (e.g. John Gamber's contribution) or compare it with novels like Ozeki's *My Year of Meats* (e.g. Bella Adams's article). I thank the reviewers for alerting me to the existence of this new book on Yamashita before sending the manuscript to press.

18. Although it can hardly be considered "nature writing" in the traditional sense, *Through the Arc of the Rain Forest* has also been read as such, specifically as an allegory of the natural cycle of the rainforest (Ishihara 1997).

19. The narrating ball can be literally and figuratively linked to the Matacão, but it can also be construed as a replica of the larger ball that is our planet. For an insightful analysis of the ball as narrator and of Yamashita's narrative strategies, see Rody (2000). In "Local Rock and Global Plastic," Heise also addresses the consequences of the choice of a nonhuman narrator (2004, 147–149). For a reading of the ball's ambiguous import, see Simal (2009, 215–217). Finally, for an interpretation of the ball as "the memory of modernity" and beyond, see Wess (2005, 112–113).

20. For a discussion of "Waste Theory," see "'The waste of the empire': Neocolonialism and Environmental Justice in Merlinda Bobis's 'The Long Siesta as a Language Primer'" (Simal Forthcoming).

21. According to the ecological theory put forward by Ellis and Ramankutty, we should favor the term "anthropogenic biomes" or *anthromes*, instead of just "biomes," in order to "describe the terrestrial biosphere in its contemporary, human-altered form, using global ecosystem units defined by global patterns of sustained direct human interaction with ecosystems, offering a new way forward for ecological research and education" (2008).

22. Interestingly, the zoo has been interpreted as an excellent illustration of "how our desire for the real can give rise to the hyperreal, to a culture in which imitations are the dominant form of reality" (Phillips 2003, 21). In zoo-like Chicolándia Yamashita takes this trope even further, as she makes this site doubly artificial by introducing plastic animals.

23. For the author, Yamashita's novel manages to "synthesize the kind of Marxist critique of postmodernism offered by Jameson with the kind of interrogation of the nature/culture binary offered by Latour" (2000, 146).

24. Retrospectively, we understand that the description of Chico Paco's improvised altar is a proleptic version of such an explanation. When trying to
fulfill his pilgrim promise and erect an altar, the boy "had gathered an
enormous amount of iron refuse, nails, aluminum cans, plastic wrappings
and plastic containers from a garbage dumb. He had filled a large base with
all this trash, which he, in turn, melted down into a solid mass with a weld-
ing torch. The combination of these materials, in fact, simulated the physical structure of the Matacão itself, creating a magnetic attraction that
proved irresistible" (97).

25. In *Through the Arc of the Rain Forest*, for instance, there is an explicit reference to "the sweat of human labor" mixing with the forest (145). See
Chap. 4 for a discussion of the interactions between human labor and the
environment in Shawn Wong and Maxine Hong Kingston.

26. The example used by Phillips a few pages later hints at the nostalgia for the
lost, original "essence" of nature as much as for the actual deterioration of
our planet: "The fiberglass fish look more like live fish than stuffed fish ever
did. Of course, such trophies memorialize much more than a great day
afield: they are monuments to a disappearing natural world" (1994, 209).

27. For an analysis of the conjunction of magic(al) realism and ecocriticism, see
Simal (2009). For the dangers of applying magical realism to Asian
American literature, see Simal (2012).

28. To substantiate his theory of the "betrayed Eden" (Buell 1998, 647), Leo
Marx includes analyses of canonical American texts from the perspective of
this "interrupted idyll" and the clash of tropes that it entails.

29. The "Rain Forest parking lot" has also modified the customs and attire of
some Amazonian Indians, who sport "reflective materials in the masks,
headpieces and necklaces" thanks to the old mirrors from the car and plane
cemetery (100).

30. With "the advent of the global," identities have been so thoroughly altered
that, as Koshy cautions us, it now becomes "imperative that the transformed meaning of the ethnic [and national] subject in transnationality be
reexamined" (2005, 111, 117). Such a reexamination, however, lies well
beyond the scope of this book.

31. In his recent "Dancing with Goblins in Plastic Jungles," Gamber claims
that rather than dismantling "the binary of the real and the fake, or the
natural and the artificial," Yamashita's novel "relies heavily on very traditional, and specifically Romantic and Transcendentalist notions of Nature"
(2018, 40). In fact, Gamber argues, Yamashita's strategy is to "underline"
rather than undermine clear-cut divisions "between the natural and the
artificial," privileging the former (2018, 49–50).

32. For earlier discussions of the impact of globalization and deterritorialization in Yamashita's work, see Wallace (2000), Heise (2004, 2008a, b), Lee
(2007), and Simal (2009, 2010).

33. Coincidentally, Gamber's new study, which focuses on the significance of birds in *Through the Arc of the Rain Forest*, also sees echoes of Carson's work in Yamashita's novel and discusses them at some length (2018, 50–51).

34. For a discussion of the dangers of biological-cultural analogies in literary texts, see Heise (2008a, 383–87, 2008b, 28–49; cf. Gamber 2012, 13).

35. For different interpretations of the type of pastoralism encoded in Yamashita's novel, see Heise (2004), Simal (2009, 2010), and Gamber (2018). In contrast to the usual reading of the final chapter as an example of "agrarian optimism" (Gamber 2018, 44) or pastoral "bliss" (Heise 2004, 138), I still maintain that the last chapter of the novel can be more fruitfully read as an example of complex (post)pastoralism or comic apocalypse (Simal 2009, 2010).

36. For an analysis of gender issues in Ozeki's novels, see Monica Chiu's "Postnational Globalization and (En)Gendered Meat Production in Ruth L. Ozeki's *My Year of Meats*" (2001), Shameem Black's "Fertile Cosmofeminism: Ruth L. Ozeki and Transnational Reproduction" (2004), and Youngsuk Chae's "'Guns Race, Meat, and Manifest Destiny': Environmental Neocolonialism and Ecofeminism in Ruth Ozeki's *My Year of Meats*" (2015). In her essay on *My Year of Meats* and *All Over Creation*, Black prefers to describe Ozeki's novels as "cosmofeminist," instead of using the ecofeminist label, arguing that the female subject position erodes the very physicality of borders and thus "stand[s] at the forefront of transnational forms of life" (2004, 226).

37. Meat symbolism and the patriarchal oppressive system are indeed linked. For Chiu, "circulating [cultural] capital becomes a euphemism for prostitution between TNCs (pimps) and the women who become vehicles of sale, promoting masculine American beef to female Japanese consumers" (2001, 106).

38. Another link between the two women, relevant not only in the plot but also in the very "literary texture" of the novel, is their common love of Sei Shonagōn's *The Pillow Book*.

39. Jane's suggestions for the TV program become more and more radical so as "to counter the underlying sexist and racist (or exaggeratedly conservative) content that Joichi promotes" (Chiu 2001, 100). For Chiu, however, Ozeki's novel is not ultimately successful, for the same "American Dream" ideology stays in place even after the ingredient of multiethnic (and other types of) diversity is introduced: "beef is coupled with American Dream families, subsequently undercut by 'real' immigrant families whose stories— advocating a 'pull-themselves-up-by-their-bootstraps' mentality—and their pig-cum-pork suppers become re-imagined fodder for overseas consumption. Essentially, nothing has been revised in order to promote a more thoughtful understanding of a diverse (and often embattled) America!" (2001, 107).

40. Those dangerous, even criminal practices include what the narrator calls the "illegal hormone ring" (358), the "pharmaceutical farming methods that taint beef and sicken humans," and the very "suffering of animals in abattoirs" (Chiu 2001, 101). In order to unveil such illegal practices, Ozeki resorts to some of the traditional conventions of detective fiction, with some differences; for instance, the monthly/menstrual chapters have clear female resonances not often encountered in classic detective fiction.

41. Andrew Wallis (2013) hailed Ozeki's first novel, *My Year of Meats*, as an accomplished example of global eco-consciousness. Ozeki herself notes that "the notion of interdependence" was pervasive and even "literal" in the first two books: "you are what you eat—the connections between the way our food is produced and who we become. And the global reach of our food networks is a playing-out of that interdependence" (Ty 2013, 162). For an analysis of the implications of the globalization of food, see Ecker (2000) and Torreiro (2016).

42. In the novel, this is confirmed by the increase in cancer cases after the consumption of genetically modified potatoes. For a specific study of Ozeki's narrative handling of GMOs, see Wallace's "Discomfort Food: Analogy, Biotechnology, and Risk in Ruth Ozeki's *All Over Creation*" (2011).

43. This strategy is not without its critics. For Heise, in trying to draw parallels between biological and cultural diversity, the novelist inadvertently conflates two realms that not always follow the same logic (2008a, 399–400).

44. For a discussion of the relationship between ecology and quantum physics, see Bohm (1980), Capra (1996), and Phillips (2003).

45. The range of matters directly or indirectly examined in the book was rather overwhelming as some reviewers pointed out. Liz Jensen, for instance, in her review for *The Guardian*, described Ozeki's *A Tale for the Time Being* as a "novel about everything from the Japanese tsunami and Silicon Valley to Zen and the meaning of life," a narrative which "sucks the reader in like a great Pacific gyre" and thus threatens to drown us if we are not careful (2013). Not only Jensen but other reviewers like Moseley highlighted that such impressive variety of topics might overwhelm the average reader.

46. A summary of the following analysis was presented at the SAAS conference held in Cáceres (Spain), in April 2017.

47. Behind the intended homophony of Nao's "now," the narrative hides an imprecise past that reaches Ruth's present in unpredictable ways. The author's namesake, Ruth, seems to function as the writer's alter ego, in an intriguing fictive-autobiographical maneuver that has been cogently analyzed by Rocio Davis in a recent article (2015).

48. For his latest eco-artistic project, the "NeoEocene site," Oliver has opted for taking "the long view" and prepares for the new scenario that global warming will bring about. He gets hold of a plot of land where he wants "to create a climate-change forest" full of "groves of ancient natives ...

species that had been indigenous to the area during the Eocene Thermal Maximum, some 55 million years ago" (60). However, conservationist policies prevent him from carrying out the NeoEocene project, because such species are not considered native to Western Canada. This triggers a fascinating discussion about how we decide what can be considered exotic, invasive species. Oliver wonders that, "given the rapid onset of climate change, we need to radically redefine the term *native* and expand it to include formerly, and even prehistorically, native species" (120). See Heise (2016, 212) for an ecocritical analysis of "'rewilding' projects."

49. "When I start writing, I start with *a sense of exploration rather than an agenda*. It's not like I have a secret plot to educate people about the evils of genetic modification or anything like that; it's more that I have a concern or a worry myself. So the book is an excuse to do the exploration" (Ty 2013, 161; emphasis added).

50. See Wallace (2016, 131–4) for an in-depth discussion of plastic materiality.

51. A brief clarification is in order. Both Harvey's project in *The Condition of Postmodernity* and the earlier work it is indebted to, Jameson's exploration of postmodernism as "the cultural dominant of the logic of late capitalism," apparently echo the Marxist base-superstructure dichotomy, for the material conditions of postmodernity seem to dictate a certain type of cultural superstructure known as postmodernism, which in turn might then be seen as a reflection of "a shift in the way capitalism is working these days" (1990, 112; 283). However, like many intellectuals of the late twentieth century, Harvey is wary of such a simplified approach to culture and, accordingly, calls attention to that risk: "it is just as surely dangerous to presuppose that postmodernism is solely mimetic rather than an aesthetic intervention in politics, economy, and social life in its own right. ... Changes in the way we imagine, think, plan, and rationalize are bound to have material consequences" (114–115).

52. By 1990, when his book was first published, Harvey believed that the new organization of capitalism had meant the effective compression of both time and space: "a strong case can be made that the history of capitalism has been characterized by speed-up in the pace of life, while so overcoming spatial barriers that the world sometimes seems to collapse inwards upon us. The time taken to traverse space ... and the way we commonly represent that fact to ourselves ... are useful indicators. ... As space appears to shrink to a 'global village' of telecommunications and a 'spaceship earth' of economic and ecological interdependencies ... and as time horizons shorten to the point where the present is all there is ..., so we have to learn how to cope with an overwhelming sense of *compression of our spatial and temporal worlds*" (240, emphasis added).

53. As Song explains, texts dealing with globalization and the changes it has brought to time-space coordinates tend to "jump from location to location, ceaselessly occupy one perspective and then another, switch between the first-person singular to a free indirect speech that bounds from character to character without respect for nationality or language, and jumble past events with present occurrences" (2011, 555). This is the case of Ozeki's novel.

54. Nao uses Google and popular webpages like Amazon and watches her own virtual funeral in the Internet.

55. An earlier version of this section was presented at the 2016 EASLCE conference, held in Brussels.

56. In fact, Harvey had already pointed at such parallelism when he had resorted to the image of the "spaceship earth" in order to capture both "economic and ecological interdependencies" (240). Even the ICTs can be viewed as indirectly fostering an ecological consciousness. In contrast to Ozeki's novel, whose depiction of ICTs is not entirely positive, in Yamashita's *Tropic of Orange*, we encounter a redeeming aspect of these technologies. In his analysis of Yamashita's novel, Gamber describes how Gabriel, one of the main characters, comes to understand the role that ICTs play in developing awareness both of "global communities" and of our interdependent, global environment: in the end, Gabriel "does come to realize that he is connected to everything else, as the digital technology behind the internet facilitates an ecological awareness" (2012, 153).

57. See Starr (2016, 112–113) for a different interpretation, according to which Nao moves away from Buddhism.

58. The quotation opening Part 3 is important in order to understand how Ozeki tries to link scientific theories with Buddhist principles. The differences between both translations are notable, and Ozeki's version clearly emphasizes the ambiguity of the phrase "time being" as a being in time, whereas other translations keep the literal meaning of "for the time being" as temporary, impermanent.

59. This interpretation can be buttressed by the fact that Ruth Ozeki, the conspicuous alter ego of Ruth the character, is herself an ordained Buddhist priest.

60. For a full description of the thought experiment, see Melody Kramer's "The Physics Behind Schrödinger's Cat Paradox" (2013).

61. What is more, the same can be applied to the observer. This interpretation, however, is parodied in the novel when Ruth watches Pesto observing his asshole: "It didn't seem like this observation caused him to split into multiple cats with multiple assholes" (Ozeki 398).

62. In a recent interview, Ozeki intimates that her inclusion of quantum mechanics emerged out of her philosophical disquisition about reality, especially the reality of characters and human beings: "Are these people real, or are they

not real? What is reality? And here is where the quantum universes come in" (Ty 2013, 162). The literary interest in such existential matters is far from new in literature: it is explicitly articulated through similar "author"-character confrontations in twentieth-century classics such as Miguel de Unamuno's novel *Niebla* (Fog) and Luigi Pirandello's play *Six Characters in Search of An Author*, published in 1914 and 1920, respectively.

63. Buddhist non-duality is invoked on a few other occasions, as when old Jiko describes the surfers and the waves as time beings, and thus not fundamentally separate (194; cf. 204). See Starr (2016, 102) for a negative appraisal of Buddhist non-dualism.

64. Ozeki concurs that "there is definitely an optimistic sensibility at the end of the book," since, "by the act of writing that final letter, Ruth is calling Nao into being in the same way that Nao called Ruth into being at the beginning of the book" (Ty 2013, 169).

65. For Rigby, "materialist ecological thought […] could be considerably enriched by entering into dialogue with older forms of nonreductive materialism" such as Buddhism (2014, 284). In their later publications, both Rigby and Gaard explore the extent to which matter is endowed with spirit, how spirituality matters. Gaard specifically discusses the ways in which Buddhism intersects with material ecocriticism. However, Buddhism is not the only religious tradition that can contribute to ecological thought and foster the care of the environment. Among indigenous traditions, eco-critics like Plumwood and Rigby point at Aboriginal spirituality as a viable way to "re-enchant" matter (see Chap. 4 for a discussion of the complex relationships of environmentalism and Native Americans). Notwithstanding the fact that Christian beliefs have been invoked to justify the exploitation of nature for centuries, there is an important Franciscan undercurrent in Christianity that has always fostered respect and love of nature and which, as Bill McKibben has recently noted (2015), has reemerged with Pope Francis and his *Laudato Si* (2015).

66. The old and new worldviews evoked by oil lamps and LEDs are also embodied by Ruth and Oliver, as they lie in bed, side by side. Of these two characters, it is Ruth whose outlook on life more clearly changes in the course of the narrative. As she becomes more engrossed in Nao's diary and learns more about Zen Buddhism, Ruth starts to question her postmodern skepticism and becomes more open to immaterial matters.

67. In reaction to the linguistic turn in theory, the proponents of material ecocriticism have urged us to question abstract concepts and focus on "the concreteness of existential experience" instead (Iovino and Oppermann 2012, 452). If constructivist theories saw "everything as textuality, as networks of signifying systems of all kinds," SueEllen Campbell urges ecocritics to look elsewhere for guidance, to learn from the ecological science, which "insists that we pay attention not to the way things have meaning for

us, but to the way the rest of the world—the nonhuman part—exists apart from us and our languages" (1996, 133).

68. Original (Latour 1991): "le trou de l'ozone est trop social et trop narré pour être vraiment naturel."

69. For Jameson, the central "antinomy of the postmodern" resides precisely in this attempt to juggle the "antifoundationalist ball," so to speak, with "the passionate ecological revival of a sense of Nature" at the same time (1994, 46–47, quoted in Wallace 2000, 138). For a study of the ways such an antinomy is overcome, see Wallace (2000); for a Burkean inflection to Wallace's discussion, see Wess (2005).

References

Adams, Rachel. 2001. The Worlding of American Studies. *American Quarterly* 53 (4): 720–732. https://doi.org/10.1353/aq.2001.0034.

Alaimo, Stacy. 2010. *Bodily Natures: Science, Environment, and the Material Self.* Bloomington: Indiana University Press.

Armbruster, Karla, and Kathleen R. Wallace, eds. 2001. *Beyond Nature Writing: Expanding the Boundaries of Ecocriticism.* Charlottesville: University Press of Virginia.

Asian American Writers' Workshop (AAWW). n.d. http://aaww.org/curation/page-turners-asian-american-literature-in-the-21st-century. Accessed Mar 1 2014.

Bahng, Aimee. 2008. Extrapolating Transnational Arcs, Excavating Imperial Legacies: The Speculative Acts of Karen Tei Yamashita's *Through the Arc of the Rain Forest. MELUS* 33 (4): 123–144.

Bartosch, Roman. 2016. Ghostly Presences: Tracing the Animal in Julia Leigh's *The Hunter.* In *Creatural Fictions*, ed. David Herman, 259–275. New York: Palgrave Macmillan.

Bate, Jonathan. 1996. Living with the Weather. *Studies in Romanticism* 35 (3): 431–447. JSTOR. http://www.jstor.org/stable/25601183

Beck, Ulrich. (1986) 1992. *Risk Society.* London: Sage.

Berry, Wendell. 1989. The Futility of Global Thinking. *Harper's Magazine*, September: 16–41.

Black, Shameem. 2004. Fertile Cosmofeminism: Ruth L. Ozeki and Transnational Reproduction. *Meridians* 5 (1): 226–256. Project MUSE.

Bohm, David. (1980) 2002. *Wholeness and the Implicate Order.* London: Routledge.

Buell, Lawrence. 1998. Toxic Discourse. *Critical Inquiry* 24 (3): 639–665.

———. 2003. Green Disputes: Nature, Culture, American(ist) Theory. In *"Nature's Nation" Revisited. American Concepts of Nature from Wonder to Ecological Crisis*, ed. Hans Bak and Walter W. Holbling, 43–50. Amsterdam: VU University Press.

————. 2005. *The Future of Environmental Criticism: Environmental Crisis and Literary Imagination*. Oxford: Blackwell.

Campbell, SueEllen. 1996. The Land and Language of Desire: Where Deep Ecology and Post-Structuralism Meet. In *The Ecocriticism Reader: Landmarks in Literary Ecology*, ed. Cheryll Glotfelty and Harold Fromm, 124–136. Athens: University of Georgia Press.

Capra, Fritjof. 1996. *The Web of Life: A New Scientific Understanding of Living Systems*. New York: Anchor Books.

Castells, Manuel. (1996) 2009. *The Information Age: Economy, Society and Culture Vol. I: The Rise of the Network Society*. Malden: Blackwell.

————. (1997) 2009. *The Information Age: Economy, Society and Culture Vol. II: The Power of Identity*. Malden: Blackwell.

————. (1998) 2010. *The Information Age: Economy, Society and Culture Vol. III: End of Millennium*. Malden: Blackwell.

Chae, Youngsuk. 2015. "'Guns Race, Meat, and Manifest Destiny': Environmental Neocolonialism and Ecofeminism in Ruth Ozeki's *My Year of Meats*." In *Asian American Literature and the Environment*, ed. Lorna Fitzsimmons, Youngsuk Chae, and Bella Adams, 126–146. New York: Routledge.

Cheung, King-kok. 1990. The Woman Warrior Versus the Chinaman Pacific: Must a Chinese American Critic Choose between Feminism and Heroism? In *Conflicts in Feminism*, ed. Marianne Hirsch and Evelyn Fox Keller, 234–251. New York: Routledge.

————, ed. 1997. *An Interethnic Companion to Asian American Literature*. Cambridge: Cambridge University Press.

Chiu, Monica. 2001. Postnational Globalization and (En)Gendered Meat Production in Ruth L. Ozeki's *My Year of Meats*. *LIT* 12 (1): 99–128.

Chuh, Kandice, and Karen Shimakawa, eds. 2001. *Orientations: Mapping Studies in the Asian Diaspora*. Durham: Duke University Press.

Clark, Timothy. 2014. Nature, Post Nature. In *The Cambridge Companion to Literature and the Environment*, ed. Louise Westling, 75–89. Cambridge: Cambridge University Press.

Crutzen, Paul J., and Eugene F. Stoermer. 2000. The 'Anthropocene.' *Global Change Newsletter* 41: 17–18.

Davis, Rocio G. 2015. Fictional Transits and Ruth Ozeki's *A Tale for the Time Being*. *Biography* 38 (1): 87–103. https://doi.org/10.1353/bio.2015.0007.

Davis, Rocio G., and Sami Ludwig, eds. 2002. *Asian American Literature in the International Context: Readings on Fiction, Poetry and Performance*. Münster: LIT Verlag.

Deitering, Cynthia. (1992) 1996. The Postnatural Novel: Toxic Consciousness in Fiction of the 1980s. In *The Ecocriticism Reader: Landmarks in Literary Ecology*, ed. Cheryll Glotfelty and Harold Fromm, 196–203. Athens: University of Georgia Press.

Ecker, Gisela. 2000. Eating Identities, from Migration to Lifestyle: Mary Antin, Ntozake Shange, Ruth Ozeki. In *Wandering Selves: Essays on Migration and Multiculturalism*, ed. Michael Porsche and Christian S. Berkemeier, 171–183. Berlin: Die Blaue Eule.

Ellis, Erle C., and Navin Ramankutty. 2008. Putting People in the Map: Anthropogenic Biomes of the World. *Frontiers in Ecology and the Environment* 6 (8): 439–447.

Flood, Alison. 2014. Ruth Ozeki Beats Thomas Pynchon to Top Kitschie Award. *The Guardian*, February 13. https://www.theguardian.com/books/2014/feb/13/ruth-ozeki-thomas-pynchon-kitschie-award

Gaard, Greta. 2014. Mindful New Materialism. In *Material Ecocriticism*, ed. Serenella Iovino and Serpil Oppermann, 291–300. Bloomington: Indiana University Press.

Gamber, John B. 2012. *Positive Pollutions and Cultural Toxins: Waste and Contamination in Contemporary U.S. Ethnic Literatures*. Lincoln: University of Nebraska Press.

———. 2018. 'Dancing with Goblins in Plastic Jungles': History, Nikkei Transnationalism, and Romantic Environmentalism in *Through The Arc of The Rain Forest*. In *Karen Tei Yamashita: Fictions of Magic and Memory*, ed. Robert A. Lee, 39–58. Honolulu: University of Hawai'i Press.

Garrard, Greg. 2004. *Ecocriticism*. London: Routledge.

———. 2011. *Ecocriticism*. 2nd ed. London: Routledge.

Gifford, Terry. 1995. *Green Voices: Understanding Contemporary Nature Poetry*. Manchester: Manchester University Press.

———. 1999. *Pastoral*. London: Routledge.

Goyal, Yogita, ed. 2017. *The Cambridge Companion to Transnational American Literature*. Cambridge: Cambridge University Press.

Haraway, Donna. (1991) 1993. A Cyborg Manifesto. In *The Cultural Studies Reader*, ed. Simon During, 271–291. New York: Routledge.

Harvey, David. 1990. *The Condition of Postmodernity*. Oxford: Blackwell.

Heise, Ursula K. 2004. Local Rock and Global Plastic: World Ecology and the Experience of Place. *Comparative Literature Studies* 41 (1): 126–152.

———. 2008a. Ecocriticism and the Transnational Turn in American Studies. *American Literary History* 20 (1–2): 381–404.

———. 2008b. *Sense of Place and Sense of Planet: The Environmental Imagination of the Global*. New York: Oxford University Press.

———. 2011. Martian Ecologies and the Future of Nature. *Twentieth-Century Literature* 57 (3–4): 447–471.

———. 2013. Globality, Difference, and the International Turn in Ecocriticism. *PMLA* 128 (3): 636–643.

———. 2016. *Imagining Extinction: The Cultural Meanings of Endangered Species*. Chicago: University of Chicago Press.

Iovino, Serenella, and Serpil Oppermann. 2012. Theorizing Material Ecocriticism: A Diptych. *Interdisciplinary Studies in Literature and Environment* 19 (3): 448–475. https://doi.org/10.1093/isle/iss087.

———, eds. 2014a. *Material Ecocriticism*. Bloomington: Indiana University Press.

———. 2014b. Introduction: Stories Come to Matter. In *Material Ecocriticism*, ed. Serenella Iovino and Serpil Oppermann, 1–20. Bloomington: Indiana University Press.

Ishihara, Toshi. 1997. Karen Tei Yamashita's Through the Arc of the Rain Forest: Nature's Text as Pilgrimage. *Studies in American Literature, The American Literature Society of Japan* 34: 59–77.

Jameson, Fredric. 1984. Postmodernism, or, the Cultural Logic of Late Capitalism. *New Left Review* 146: 53–92.

———. 1991. *Postmodernism, or the Cultural Logic of Late Capitalism*. Durham: Duke University Press.

Jay, Paul. 2010. *Global Matters: The Transnational Turn in Literary Studies*. Ithaca: Cornell University Press.

Koshy, Susan. 2005. The Postmodern Subaltern: Globalization Theory and the Subject of Ethnic, Area, and Postcolonial Studies. In *Minor Transnationalism*, ed. Françoise Lionnet, and Shih Shu-mei, 109–131. Durham: Duke University Press.

Kramer, Melody. 2013. The Physics Behind Schrödinger's Cat Paradox. *National Geographic*. https://news.nationalgeographic.com/news/2013/08/130812-physics-schrodinger-erwin-google-doodle-cat-paradox-science/

Ladino, Jennifer K. 2012. *Reclaiming Nostalgia: Longing for Nature in American Literature*. Charlottesville: University of Virginia Pres.

Latour, Bruno. 1991. *Nous n'avons jamais été modernes: Essai d'anthropologie symétrique*. Paris: La Découverte.

———. 1993. *We Have Never Been Modern*. Trans. Catherine Porter. Cambridge, MA: Harvard University Press.

———. 2002. *War of the Worlds: What About Peace?* Chicago: Prickly Paradigm Press.

Lee, Sue-Im. 2007. 'We Are Not the World': Global Village, Universalism, and Karen Tei Yamashita's *Tropic of Orange*. *Modern Fiction Studies* 53 (3): 501–527.

Lee, A. Robert, ed. 2018. *Karen Tei Yamashita: Fictions of Magic and Memory*. Honolulu: University of Hawai'i Press.

Li, David Leiwei. 1998. *Imagining the Nation: Asian American Literature and Cultural Consent Literatures*. Stanford: Stanford University Press.

Ling, Jinqi. 2006. Forging a North-South Perspective: Nikkei Migration in Karen Tei Yamashita's Novels. *Amerasia Journal* 32 (3): 1–22.

Love, Glen A. 1999. Ecocriticism and Science Consilience? *New Literary History* 30 (3): 561–576.

Lowe, Lisa. 1991. Heterogeneity, Hybridity, Multiplicity: Marking Asian American Differences. *Diaspora* 1 (1): 24–44.

———. 1996. *Immigrant Acts: On Asian American Cultural Politics*. Durham: Duke University Press.

Ma, Sheng-Mei. 1998. *Immigrant Subjectivities in Asian American and Asian Diaspora Literatures*. Albany: State University of New York Press.

Marx, Leo. 1964. *The Machine in the Garden: Technology and the Pastoral Ideal in America*. New York: Oxford University Press.

———. 2002. The Pandering Landscape: On American Nature as Illusion. In *"Nature's Nation" Revisited. American Concepts of Nature from Wonder to Ecological Crisis*, ed. Hans Bak and Walter W. Holbling, 30–42. Amsterdam: VU University Press.

Marzec, Robert P. 2007. *An Ecological and Postcolonial Study of Literature: From Daniel Defoe to Salman Rushdie*. New York: Palgrave Macmillan.

Mazel, David. 2000. *American Literary Environmentalism*. Athens: University of Georgia Press.

McKibben, Bill. (1989) 2006. *The End of Nature*. New York: Random House.

———. 2015. The Pope and the Planet. *The New York Review of Books*, August 13. http://www.nybooks.com/articles/archives/2015/aug/13/pope-and-planet/

McLuhan, Marshall, and Bruce R. Powers. 1989. *The Global Village: Transformations in World Life and Media in the 21st Century*. New York: Oxford University Press.

Morton, Timothy. 2007. *Ecology Without Nature: Rethinking Environmental Aesthetics*. Cambridge: Harvard University Press.

Murphy, Patrick. 2009. *Ecocritical Explorations in Literary and Cultural Studies: Fences, Boundaries, and Fields*. Lanham: Lexington Books.

Nguyen, Viet Thanh. 2002. *Race and Resistance: Literature and Politics in Asian America*. New York: Oxford University Press.

Nicolescu, Basarab. 2007. Transdisciplinarity as Methodological Framework for Going Beyond the Science-Religion Debate. *Transdisciplinarity in Science and Religion* 2: 35–60.

Ortiz-Robles, Mario. 2016. *Literature and Animal Studies*. New York: Routledge.

Ozeki, Ruth. 1998. *My Year of Meats*. New York: Penguin.

———. 2003. *All Over Creation*. New York: Viking.

———. 2013. *A Tale for the Time Being*. Edinburgh: Canongate.

Palumbo-Liu, David. 1999. *Asian/American: Historical Crossings of a Racial Frontier*. Stanford: Stanford University Press.

Parikh, Crystal, and Daniel Y. Kim. 2015. *The Cambridge Companion to Asian American Literature*. New York: Cambridge University Press.

Phillips, Dana. 1994. Is Nature Necessary?. In *The Ecocriticism Reader: Landmarks in Literary Ecology*, ed. Cheryll Glotfelty and Harold Fromm, 204–222. Athens: University of Georgia Press.

————. 1999. Ecocriticism, Literary Theory, and the Truth of Ecology. *New Literary History* 30 (3): 577–602.

————. 2003. *The Truth of Ecology: Nature, Culture, and Literature in America.* New York: Oxford University Press.

Pope Francis. 2015. *Laudato Si: On Care for Our Common Home.* London: Catholic Truth Society.

Rigby, Kate. 2014. Spirits That Matter: Pathways Towards a Rematerialization of Religion and Ecospirituality. In *Material Ecocriticism*, ed. Serenella Iovino and Serpil Oppermann, 283–290. Bloomington: Indiana University Press.

————. 2015. Ecocriticism. In *Introducing Criticism at the Twenty-First Century*, ed. Julian Wolfreys, 2nd ed., 122–154. Edinburgh: Edinburgh University Press.

Rody, Caroline. 2000. Impossible Voices: Ethnic Postmodern Narration in Toni Morrison's *Jazz* and Karen Tei Yamashita's *Through the Arc of the Rain Forest.* *Contemporary Literature* 41 (4): 618–641.

Safran, William. 1991. Diasporas in Modern Societies: Myths of Homeland and Return. *Diaspora* 1 (1): 83–99.

Scholte, Jan Aart. 2005. *Globalization: A Critical Introduction.* 2nd ed. London: Palgrave Macmillan.

Shan, Te-Hsing. 2006. Interview with Karen Tei Yamashita. *Amerasia Journal* 32 (3): 123–142.

Simal, Begoña. 2009. Of a Magical Nature: The Environmental Unconscious. In *Uncertain Mirrors: Magical Realism in US Ethnic Literatures*, ed. Jesús Benito, Ana Manzanas, and Begoña Simal, 181–224. New York: Rodopi.

————. 2010. The Junkyard in the Jungle: Transnational, Transnatural Nature in Karen Tei Yamashita's *Through the Arc of the Rain Forest.* *Journal of Transnational American Studies* 2 (1): 1–25.

————. 2012. Of Magical Gourds and Secret Senses: The Uses of Magical Realism in Asian American Literature. In *Moments of Magical Realism in US Ethnic Literatures*, ed. Lyn Di Iorio Sandín and Richard Perez, 123–152. New York: Palgrave Macmillan.

————. 2013. Judging the Book by Its Cover: Phantom Asian America in Monique Truong's *Bitter in the Mouth. Concentric* 39 (2): 7–32.

————. 2018. Introduction to Special Forum 'Disrupting Globalization: Transnationalism and American Literature'. *Journal of Transnational American Studies* 9 (1): 277–291.

————. Forthcoming. 'The Waste of the Empire': Neocolonialism and Environmental Justice in Merlinda Bobis's 'The Long Siesta as a Language Primer'. *The Journal of Postcolonial Writing* 55 (2019).

Simal, Begoña, and Elisabetta Marino, eds. 2004. *Transnational, National, and Personal Voices: New Perspectives on Asian American and Asian Diasporic Women Writers.* Münster: LIT Verlag.

Sohn, Stephen H., Paul Lai, and Donald C. Goellnicht. 2010. Introduction: Theorizing Asian American Fiction. *Modern Fiction Studies* 56 (1): 1–18.

Song, Min Hyoung. 2011. Becoming Planetary. *American Literary History* 23 (3): 555–573.

Soper, Kate. 1998. *What Is Nature? Culture, Politics, and the Non-Human.* Malden: Blackwell.

Starr, Marlo. 2016. Beyond Machine Dreams: Zen, Cyber, and Transnational Feminisms in Ruth Ozeki's *A Tale for the Time Being. Meridians* 13 (2): 99–122. https://doi.org/10.2979/meridians.13.2.06.

Stiglitz, Joseph. 2017. *Globalization and Its Discontents Revisited: Anti-Globalization in the Era of Trump.* London: Penguin.

Torreiro, Paula. 2016. *Diasporic Tastescapes: Intersections of Food and Identity in Asian American Literature.* Zurich: LIT Verlag.

Tuana, Nancy. 2008. Viscous Porosity: Witnessing Katrina. In *Material Feminisms,* ed. Stacy Alaimo and Susan Hekman, 188–213. Bloomington: Indiana University Press.

Ty, Eleanor. 2013. 'A Universe of Many Worlds': An Interview with Ruth Ozeki. *MELUS* 38 (3): 160–171. https://doi.org/10.1093/melus/mlt028.

Wallace, Molly. 2000. 'A Bizarre Ecology': The Nature of Denatured Nature. *Interdisciplinary Studies in Literature and Environment* 7 (2): 137–153.

———. 2011. Discomfort Food: Analogy, Biotechnology, and Risk in Ruth Ozeki's *All Over Creation. Arizona Quarterly* 67 (4): 135–161.

———. 2016. *Risk Criticism: Precautionary Reading in an Age of Environmental Uncertainty.* Ann Arbor: University of Michigan Press.

Wallis, Andrew H. 2013. Towards a Global Eco-Consciousness in Ruth Ozeki's My Year of Meats. *Interdisciplinary Studies in Literature and Environment* 20 (4): 837–854. https://doi.org/10.1093/isle/ist088.

Weil, Kari. 2010. A Report on the Animal Turn. *differences* 21 (2): 1–23. https://doi.org/10.1215/10407391-2010-001.

Wess, Robert. 2005. Terministic Screens and Ecological Foundations: A Burkean Perspective on Yamashita's *Through the Arc of the Rain Forest. Interdisciplinary Literary Studies: A Journal of Criticism and Theory* 7 (1): 104–115.

Williams, Raymond. 1980. Ideas of Nature. In *Culture and Materialism: Selected Essays,* 67–85. London: Verso.

Wong, Sau-Ling Cynthia. 1993. *Reading Asian American Literature: From Necessity to Extravagance.* Princeton: Princeton University Press.

———. 1995. Denationalization Reconsidered: Asian American Cultural Criticism at a Theoretical Crossroads. *Amerasia Journal* 21 (1–2): 1–27.

Yamashita, Karen Tei. 1990. *Through the Arc of the Rain Forest.* Minneapolis: Coffee House Press.

———. 2010. *I Hotel.* Minneapolis: Coffee House Press.

Zavarzadeh, Mas'ud, and Donald Morton. 1991. *Theory, Pedagogy, Politics: The Crisis of the 'Subject' in the Humanities.* Urbana: University of Illinois Press.

CHAPTER 7

Coda: Weedflowers, Gold Mountains, and Murky Globes

When the late Amy Ling and Annette White-Parks, probably inspired by Edith Eaton's recurring vegetal imagery, opened their 1995 edition of *Mrs. Spring Fragrance and Other Writings* with the description of the burgeoning field of Asian American literature, they chose the quintessential natural metaphor: a flowering, "many-branched" tree, "coming into fuller and fuller bloom" after the first scholarly "shoots" of the 1970s and 1980s, a tree that had also proved "to be deeply rooted," dating back to the nineteenth century (1995, 1). In our ecocritical journey, we have gone back to those inaugural moments of Asian American literature, with Eaton's pioneering work and with the first immigrant encounters with the new land; we have also revisited master narratives like the Japanese American internment and the claiming of America. Needless to say, we have reexamined those old sites with new eyes, in this case through a green lens. Heeding Elaine Kim's advice to go "beyond railroads and internment" (1995, 12–13), we have also engaged with more recent Asian American writings, which have brought new concepts and new formal strategies, with their speaking miniature planets (Yamashita) and ominous "murky globes" (Ozeki).

The new ideas and images so skillfully wielded by Ozeki and Yamashita, however, should not make us forget older concepts and discourses, like the pastoral mode, which continue to be valuable when appropriated and transformed into critical (post)pastoralism. We can find literary jewels not only in old ecocritical discourses but also in the tradition of Asian American literary criticism itself. In 1993, Sau-ling Wong published her landmark

© The Author(s) 2020   267
B. Simal-González, *Ecocriticism and Asian American Literature*,
Literatures, Cultures, and the Environment,
https://doi.org/10.1007/978-3-030-35618-7_7

Reading Asian American Literature. In her introduction to the book, Wong tried to build "a sense of an internally meaningful literary tradition" (1993, 11) and two axes or "riverbanks" guided her in that inspired task, Necessity and Extravagance. These she defined as "two contrasting modes of existence and operation, one contained, survival-driven and conservation-minded, the other attracted to freedom, excess, emotional expressiveness, and autotelism" (13). While Necessity was usually associated with the frugal attitudes of first-generation immigrants, whose motto was "no waste," the members of subsequent generations, American-born and raised, were often tempted by the lures of Extravagance and, in the eyes of their parents or elders, squandered both money and energy. They wanted to "fly high kites," as Kingston would put it. I foresee that the waste/no-waste dialectics that underpins Wong's Necessity/Extravagance binary will continue to be useful in the future. This pervasive dichotomy, which, as Wong argued, erected and consolidated intergenerational walls in many Asian American texts during the twentieth century, is still present in recent immigration narratives, where it continues to signify the rift between American(-born) and immigrant characters. In our transnational context, the Necessity/Extravagance dynamics has acquired a new dimension, since it is often deployed in order to underscore the distance between diasporic subjects and those who stayed behind, in the various Asian homelands. In analyzing the relationship between Asian American literature and the environment, as I have attempted to do in this book, that old dichotomy offers an unexpected gift, for it can now be read in ecological terms as the dialectics of austerity versus excess, degrowth versus consumerism. Although the older generation's command not to waste was not intended as an environmentalist injunction in its original (con)texts, an ecocritical reading of such a discourse unearths its environmental potential, a new, productive field which will certainly invite future investigations.

As my own journey of ecocritical exploration draws to an end, I notice that we have traveled a long way, from the initial first contacts with natural and toxic landscapes to Ozeki's full "array of possibilities." In the meantime, I have examined the ways in which Asian American literature digests (weed)flowers, "storied mountains" (Hayashi 2007, 1), plastic flesh, and even miniature globes. The fact that Yamashita has chosen to give voice to the planet in her novel indicates a new ecological awareness. In addition, the fact that the small globes in recent Asian American literature adopt the shape of crystal or plastic simulacra of the real, larger globe says something of the current state of our planet. These disintegrating voices of the earth, these "murky globes," point at the anthropogenic degradation of our planet and underscore the end of nature as we used to know it.

Not everything is lost yet. If Lawrence Buell is right when he claims that "the environmental crisis" is inextricably linked with "a crisis of the imagination" (1995, 2), we can entertain some hope, for our literary imagination is far from dead. There is an increasing body of imaginative literature that, while often reminding us of our material limits, also points at an "array of possibilities," a multitude of ways in which we can, ecologically, move forward.

REFERENCES

Buell, Lawrence. 1995. *The Environmental Imagination: Thoreau, Nature Writing, and the Formation of American Culture*. Cambridge, MA: Harvard University Press.

Hayashi, Robert T. 2007. *Haunted by Waters: A Journey Through Race and Place in the American West*. Iowa City: University of Iowa Press.

Kim, Elaine. 1995. Beyond Railroads and Internment: Comments on the Past, Present, and Future of Asian American Studies. In *Privileging Positions: The Sites of Asian American Studies*, ed. Gary Y. Okihiro et al., 1–9. Pullman: Washington State University.

Ling, Amy, and Annette White-Parks. 1995. Introduction. In *Mrs. Spring Fragrance and Other Writings*, ed. Amy Ling and Annette White-Parks. 1–8. Chicago: University of Illinois Press.

Wong, Sau-Ling Cynthia. 1993. *Reading Asian American Literature: From Necessity to Extravagance*. Princeton: Princeton University Press.

INDEX[1]

---

[1] Note: Page numbers followed by 'n' refer to notes.

Printed in the United States
By Bookmasters